THE GEO. S. LONG PUBLICATION SERIES

RENEWABLE RESOURCE MANAGEMENT FOR FORESTRY AND AGRICULTURE

Edited by James S. Bethel
and Martin A. Massengale

UNIVERSITY OF WASHINGTON PRESS
Seattle and London

Printed in the United States of America

Library of Congress Cataloging in Publication Data

Renewable Resource Management for Agriculture and
Forestry Symposium, Denver, 1977.
Renewable resource management for forestry and
agriculture.

(The Geo. S. Long publication series)
Includes bibliographies and index.
1. Agricultural conservation—Congresses.
2. Forest conservation—Congresses. 3. Agriculture
—Congresses. 4. Forests and forestry—Congresses.
5. Renewable natural resources—Congresses.
I. Bethel, James Samuel. II. Massengale,
Martin Andrew, 1933- III. Title. IV. Series.
S604.5.R46 1977 338.1 78-25994
ISBN 0-295-95624-0

THE GEO. S. LONG PUBLICATION SERIES

The Geo. S. Long Fund was established in 1975 to promote a better understanding of forestry, natural resources, and conservation. The endowment was made by Miss Helen Long of Tacoma to the University of Washington in memory of her father, a distinguished timber industry executive and natural resources conservationist, who opened the first Weyerhaeuser office in Tacoma in 1900 and served as the firm's manager, vice president, and chairman of the executive committee.

The generous financial support provides an income to the University of Washington to carry out the intent of the founder with provisions that allow the dean of the College of Forest Resources to administer a program of lectures, publications, travel, and library acquisition.

The establishment of the lecture series with the introduction of distinguished authorities to the academic and forestry community has been endorsed by Miss Long as a principal effort of the fund. Publication of the Long Lectures as well as the work of other educators and scientists in associated disciplines will supplement and extend the influence of the program.

The terms of the gift are broad enough to allow the university to pursue scholarly excellence in such areas as special collections and the attendance of faculty at academic and professional meetings of special merit.

Seventy-five years elapsed between the time of Geo. S. Long's first arrival in the Pacific Northwest and the establishment of this series. With Miss Long's authorization and support the university will continue the legacy to work for a better understanding of forestry, natural resources, and conservation through academic efforts, a distinguished lecture series, and books such as this one.

Contents

Preface

A symposium on renewable resource management for forestry and agriculture was organized as a part of the 1977 annual meeting of the American Association for the Advancement of Science. The symposium represented a first major effort to encourage agricultural scientists and forestry scientists to examine jointly their common problems in the management of renewable resources. Major studies by the National Academy of Sciences, conducted by its World Food and Nutrition Steering Committee and its Committee on Renewable Resources for Industrial Materials, have addressed important food and fiber questions separately. This symposium has undertaken to address many of these same questions from the same platform.

Requirements for food and fiber, the renewable resources, are increasing rapidly worldwide. These increases are related to the growth of the human population, depletion of readily accessible nonrenewable resources, and demands from the poorer segments of the population for a better life. If there is to be a positive response to these growing needs, it will be achieved through more effective allocation and more efficient use of productive land.

In the competition for the use of land, agriculture has usually been able to displace forestry. But both forestry and agriculture have often been displaced when land was required for urban development or for roads, railroads, or reservoirs.

As human societies have developed increasingly complex organizational structures, they have imposed ever greater demands upon limited nonrenewable resources. The reserves of minerals and fossil fuels most accessible to human populations have been exhausted first, and we are now exploiting those resources that are less and less accessible—those geographically remote or located at great distances beneath the surface of the earth or the oceans. We do not know the absolute magnitude of the earth's nonrenewable material reserves. From the standpoint of meeting human need for materials this may not be important. We do know that the costs in capital, labor, and energy associated with exploration, discovery, and development of new reserves are increasing rapidly. As these costs continue to increase we approach the point where the return in material value does not justify the investment of scarce resources. At this point we have effectively exhausted our nonrenewable resources, even though large reserves are known to exist. It is the prospect of this kind of resource exhaustion that leads to a growing interest in substitution of renewable for nonrenewable materials. The substitution of renewable wood for nonrenewable metals and ceramics represents significant conservation of process energy. Essentially every petrochemical confronts technically feasible substitutes among the silvachemicals. Renewable materials are produced from the soil, the same resource that is basic to the production of food.

Human society needs both food and fiber. The land base available to agriculture and forestry is adequate to provide massive increases in both types of commodities, provided that productive land is efficiently managed for continuous production of the product allocated to it. It is most timely then for agriculturists and foresters to examine jointly the problems associated with renewable resource management. It is important that they share common problems and transfer appropriate technologies. This symposium represents an effort to develop a meaningful dialogue between the agricultural community and the forestry community in the domain of resource management. The ideas presented in this collection of symposium papers should provide the basis for a continuing dialogue across the disciplines relevant to renewable resources.

RENEWABLE RESOURCE MANAGEMENT FOR FORESTRY AND AGRICULTURE

1 Meeting world requirements for food and fiber

Emery N. Castle

THE STATE OF OUR KNOWLEDGE

In the past decade the mood in the United States regarding the availability of natural resources has shifted all the way from optimism (or complacency) to pessimism (or uncertainty), and with this change has come the realization that the United States, short of a major change in social organization, is truly dependent on the rest of the world. Our natural resource policies and programs increasingly reflect this realization—with the embarrassing exception of the theory of energy independence.

It is appropriate that agriculture and forestry be considered simultaneously in this context, not so much because both deal with renewable resources but because U.S. agriculture and forestry have a similar comparative advantage in relation to the rest of the world.[1] One has been transformed by a technological revolution and the other appears to be on the verge of substantial technological change.

Given world interdependence and the technological changes in agriculture that have required considerable investment, and given certain humanitarian concerns, it is logical to raise questions about meeting world food and fiber requirements. Several efforts have been made—using different approaches—to estimate food and fiber needs and our capacity to meet them. It is useful to ascertain what has been learned from these efforts, even though the results are divergent and the policy implications inconsistent.

Commodity-specific studies select a particular commodity, say food, and attempt either to analyze historical trends or to project requirements, or both. Global requirements are estimated to show the magnitude of the supply task or to make judgments regarding the possibilities of meeting these requirements (University of California, 1974; U.S. Department of Agriculture, 1976).

Studies of a more comprehensive nature seek to assess the demand-supply situation for many resources simultaneously and to examine the relationships between different resources in both use and production. These studies attempt to understand the aggregate impact of mass consumption and industrialization on global resources. There is merit in differentiating between studies that emphasize historical trends and those that utilize future projections, even though these classifications are not mutually exclusive. My example here of a study based on historical experience is one by Barnett and Morse (1963) which deals with trends in natural resource availability as measured by the associated resource requirements of labor and capital. The work by Meadows and others (1972) offers an example of the use of systems dynamics to predict future trends.

What has been learned from these quantitative studies of our history and future? Human consumption of all goods and services, even a basic necessity such as food, is highly variable. There is a minimum caloric and nutritional level below which life is not sustained, but requirements cannot be stated accurately on a global basis. In the first place, the nutritional knowledge base is not firm. A slight difference in assumption with respect to minimum protein needs can make a tremendous difference in resources required for food consumption. The possibilities of food substitution for those living above the subsistence level are enormous, and as income rises food satisfies many human desires beyond mere sustenance. But as income falls food becomes a residual claimant on the land and the possibilities for substitution decrease.

On the supply side, there are additional difficulties. Accurate inventories of resources do not exist. We do not know the rate at which technology will develop and be applied, nor do we know all the possibilities for substitution in production (Hayami and Ruttan, 1970).

Although forestry and agriculture both tend to be residual land claimants, with agriculture usually able to attract land away from forestry, during a period of great need they would take precedence and much land currently used for other purposes could be converted to food and fiber production. But the unpredictability of the yields and the cost of making these adjustments simply add to the complexity of estimating consumption-production possibilities for the near future, to say nothing of a longer period. For example, if all the human labor, fertilizer, and mechanical power used for home beautification were devoted to food production, and if the resources used to feed pets were made available for human consumption, the effect would be significant.

Comprehensive commodity analyses also encounter conceptual difficulties. Barnett and Morse provided important insights into trends in the use of resources by industrial and industrializing societies, even though their analysis was confined to the United States, but they were forced to rely on data of questionable quality and their conclusions are based on subjective evaluation of the data rather than statistical analysis and testing. Moreover, the period under study saw an unprecedented growth in knowledge and technology. It was a period in which the related resource requirements of resource-dependent commodities, including food, were stable or falling; yet the social rate of return has been high indeed on investments in agricultural technology in the public sector. (There is a substantial body of literature on the measurement of the social rate of return on investment in agricultural research. For example, see Griliches, 1958; Griliches, 1964; and Bredahl and Peterson, 1976.) If the production function for agricultural technology is such that diminishing marginal returns are being experienced, the resources-technology relationship obviously will change (Committee on Agricultural Production Efficiency, 1975). There is a need for economists to integrate these two kinds of studies.

Barnett and Morse found generally stable or falling relative prices except for timber, as measured by sawlog prices. A recent study by Manthy (1977) updates the Barnett-Morse work and extends it to include more than sawlogs. Manthy reports that sawlogs have been and continue to be a scarce resource in the United States. However, when Manthy extends the Barnett-Morse analysis to all forest products, he finds that this composite good is not unique and also demonstrates falling resource-related requirements. He shows that since the early 1920's the composite "price" index for forest products has declined more than the price for agricultural products and minerals. The Manthy work suggests, even in the face of modest technological change for forestry, that if all forest products are considered, they have not become relatively more scarce.[2]

The systems dynamics approach to resource availability has a superficial appeal. By providing for interdependencies and accounting for time, it can, in theory, study interactions that are frequently neglected, such as relationships between resources and technology. In practice, of course, the difficulties are formidable. Little is known about such relationships, and the introduction of time as a determinant creates an even greater need for knowledge about social learning, social incentives, and social control mechanisms as they impinge on production and consumption decisions. Comprehensive systems dynamics models typically have either ignored these relationships or have incorporated them only sparingly. As a consequence, there is a real question whether the results should be taken seriously as predictions. It has been argued that they may have value in demonstrating the consequences of a possible set of group activities.[3]

What can be generalized from these studies with respect to renewable resources (food and fiber) in particular and natural resources in general? The world's population will generate great need for foodstuffs before the end of the century. Indeed, pressure will be placed on the resource base, the knowledge base, and the food processing system to meet these expectations. The global situation with respect to fiber generally and the output of the forests in particular is somewhat less clear. Nevertheless, the trends toward deforestation, the increasing demands for fiber which accompany growth in income and industrialization, and the relatively modest probable global improvements in technology also suggest a significantly greater need for fiber on a global basis.

THE MISSING INGREDIENT

My discussion to this point has emphasized supply and technology. This is the traditional approach and there is much to be learned from such a systematic assessment of the problem. Given reasonable assumptions of population and income growth, only by an increase in output of food and fiber can a decline in standard of living be prevented.

An exclusive reliance on such an approach, however, will confront mankind with a nasty dilemma. On the one hand, there is a potential demand, for which output is likely to be inadequate unless more resources and an improved technology are put to use; on the other, the increase in output, resulting from centralized decision making, may well generate even greater demand. Unless the relations between supply and demand are perceived correctly in centralized and decentralized decisions, we will precipitate a future crisis. Thus the central issue is not whether technology can somehow be harnessed to increase food and fiber output. Of course it can. But will it be accompanied by appropriate individual decisions regarding size of family and levels of consumption of resource-using commodities? If these adjustments are not made, the inevitable result will be either centralized decision making, with a loss of individual freedom, or a confrontation with the minimum in maintenance of human life. Yet the quality of life will decline sharply long before the ultimate limits of survival are reached.

It is unclear whether the greatest threat to the quality of life will come from the undesirable side effects of widespread application of output-increasing technology, from overpopulation, or from declining per capita consumption. In the context of this study it does not matter. My own judgment is that it will come from the increasing scarcity of common-property resources as a result of mass production and consumption. I have attempted to demonstrate that increased production of food and fiber will not address the issue that must be addressed: production and consumption decisions must be coordinated globally.

Communal and tribal societies always adjust their level of living to available resources. If a society is to survive, the cause-and-effect relationships affecting human life must be reflected in the decisions made by the society. In practice, societies develop many formal and informal mechanisms to regulate and control behavior. It is not necessary for survival that there be an equilibrium between consumption and production, but it is necessary that new information be put to use by the formal and informal social control mechanisms which bring about adjustment.

According to Dunn (1971, 1975), most of the adaptations made by species to environmental challenges in the course of evolutionary history have been specializing. That is, they produce highly specialized forms of behavior within the limits of a restricted environment. Adaptive generalizations occur less frequently and widen the environmental range within which the species operates. The unique feature of the genus *Homo* is that it has made a series of generalizing adaptations.

Dunn says that human specialization did not come about because of pressures exerted by the natural environment but has developed through learning processes which have amplified human productivity. As examples of adaptive generalizations he cites the development of organized markets (cities) and of institutional science allied with technology.

The awesome capacity for growth that is a consequence of specialized and generalized adaptations has encountered new limits (Hitch, 1976). Environmental side effects are of global significance. The concentration of power affecting human welfare is enormous because of nuclear and comparable developments. The specter of an inadequate food supply results from a success of adaptations that have made increased food production possible but a failure of adaptations in consumption.

Let us reflect for a moment. There is evidence that the average nutritional level, in both quantity and quality, is higher than ever before. We also know that these improvements affect a human population larger than ever before. Thus more food is being produced for human consumption, transported greater distances, and processed to a higher level nutritionally than at any time in human history. Yet the National Academy of Sciences has completed a study designed to answer the question, "How can the United States better use its research capability to meet world food and nutrition needs?" The World Food and Nutrition Steering Committee, under the chairmanship of Dr. Harrison Brown, recommended a greatly expanded, publicly financed research program (National Academy of Sciences, 1977a). Thus we face the prospect of additional food deficits, and are girding our loins for an even more ambitious attempt to increase production by developing technology through public investment.

Let us assume that this research succeeds in doing what the scientists say it can do. That is to say, assume that we increase global food production by about 50 percent in the next 15 years and even more than that subsequently. How will this be viewed by the millions around the world who make the crucial decisions on family size, education, human health, production, harvesting, and distribution? Will it be seen as an opportunity for even larger populations and even greater consumption of material "things"? Or will the increased output inspire appropriate incentives for future adjustments?

Obviously, social adaptations and social control mechanisms are needed on a global basis to bring individual behavior into alignment with social objectives. These mechanisms have been modified indirectly by natural science through the widespread adaptation of technology, and they are a result of social experimentation and learning that stems from technical change. They have not changed to a significant extent, I believe, because of advances in knowledge in the social and behavioral sciences, although Adam Smith, Karl Marx, and J. M. Keynes might be cited to the contrary. However, currently there is much interest in the social sciences because of the obvious need to understand the origin and development of social control mechanisms.

One of the study teams working under the direction of the Steering Committee chaired by Harrison Brown has developed interesting material in this connection (National Academy of Sciences, 1977c; pp. 86–88). The team studied the development of quality-of-life indicators such as life expectancy, educational attainment, infant mortality, and per capita GNP, and compared nations and states with respect to their accomplishments as measured by these indicators. A positive correlation exists between per capita GNP and other quality-of-life indicators. Economic growth, as conventionally measured, apparently becomes a means as well as an end in itself.

But this is not universally true. There are examples of countries in the middle- and low-income groups which have achieved quality-of-life standards, except per capita GNP, comparable to the richest countries. An understanding of how this came about would help others achieve a desired quality of life without the high consumption of resources typical of industrialized societies.

There has been much discussion recently about the desirability of tailoring advances in knowledge and technology to the particular needs of the small farmer who may be operating in a barter economy. Reference is often made to the advances possible through the development of improved seeds. These seeds can be used by farmers with almost any level of technology. Of course, the productivity of fertilizers, herbicides, and pesticides may be increased when they are used in conjunction with improved seeds; nevertheless, these associated inputs are not necessary to take advantage of the improved seeds. It has been advanced that this is an example of small-farmer technology.

But is it? If the improved seeds are successful in producing a surplus on the farm, the excess will go into some kind of off-farm marketing system. To date, the most efficient of these systems have depended on modern transportation and communication. This means large systems. If the productivity of purchased inputs, such as fertilizer, is improved, incentive exists for the development of an off-farm fertilizer industry. Again, a complex, large system. Although the initial adapter may have been the small farmer, the indirect effects may not work to perpetuate small-scale operations.

Labor tends to be abundant in the developing societies and especially so on the land, but we should not forget that the tasks of farming under primitive conditions are onerous and burdensome. Even if the cost of labor is zero, there will always be pressure to find a substitute because of the disutility of labor. Indeed, this is one of the first ways that quality of life can be improved for the peasant or the primitive farmer.

The first application of an advance in knowledge may be a specializing adaptation, that is, a better adaptation to existing environmental conditions. However, generalizing adaptations must be made in order to reap the full advantages of knowledge. Thus I doubt that technological development based on the premise that "small is beautiful" will succeed even if it is possible to incorporate it at the first stage of development.

The crux of many problems is the relation between the adapter of new technology and the generalized adaptation, the large commercial system. It is unlikely that studies at the micro level, within the developing societies, will be meaningful in a policy context unless they are undertaken in a realistic macro framework. It is irresponsible to develop and promote technology abroad in the absence of explicit recognition of the problems of the industrialized world.

It is beyond the scope of this chapter to identify and discuss the problems of the industrialized societies based on successful generalized adaptations. But a major problem which can be stated is whether the political, economic, and social ramifications of complex commercial systems can be sufficiently well understood by the populace they serve to be properly managed by that populace in nontotalitarian societies. The concern about the price and supply of energy and the distribution of income provides a contemporary example in the United States.

The market is an obvious link between commercial systems and agrarian development. The market's role in providing this bridge needs to be studied by anthropologists, political scientists, and sociologists who appreciate its power to change the standard of living in developing societies. Yet the limits of the market also need to be understood. The management of commercial organizations becomes one of the principal policy issues in nontotalitarian societies. The kind of investigation needed would focus on social control mechanisms, in addition to the market, which accomplish the following: (1) cause information to be generated which is predictive of the consequences of both centralized and decentralized decisions, and (2) provide explicit and automatic means for reconciling differences among decision makers with respect to what the information portends.

The crux of the problem is whether this can be accomplished by large and complex societies without recourse to totalitarian means. Adam Smith's "invisible hand" was an excellent example of socially desirable suboptimization. His views had great historical validity and continue to be valid for many problems and situations. However, institutions, arising from successful generalizing adaptations, acquire objectives and lives of their own which do not necessarily serve the greater good with the passage of time. This applies to institutions in the private sector as well as to the public sector bureaucracies.

A GLOBAL FOOD AND FIBER POLICY

In this section I shall identify what I believe to be a desirable resource policy. These proposals are necessarily general because the scope is global and the future uncertain.

1. Public investment in research to increase the efficiency of food and fiber production is justified both in the United States and abroad. The evidence on the social rate of return to past investments of the public sector, as well as the measurements of probable future need, suggests that such investment is appropriate.

The United States is apparently on the verge of making additional significant investments in such research. This may be because it is in the self-interest of large organizations in the United States, such as those in the research establishment, commercial agriculture, and commercial forestry, to move in that direction. Nevertheless, this appears to be a case in which self-interest is consistent with a larger social good.

2. Public policies and programs are desirable which provide strong incentives to the producers of food and fiber. It is unfortunate that policies in the developing societies often tend to have the opposite effect. Cheap food is politically popular, and the public perception of the problem is such that price manipulation to favor consumers in the short run often occurs. Yet worldwide experience shows that in the long run the cheapest food is produced in response to market incentives.

Before being overly critical of the developing societies for policies which favor consumers rather than producers, and which prevent prices from serving as long-run incentives, we should note that this has often been the energy policy of the United States, especially since 1972.

3. International trade in agricultural commodities and inputs should also be facilitated. Such markets provide the principal method of integrating the economies of the developing and the more developed societies. The subsidized export of agricultural products is to be avoided, except in emergencies, because of its disrupting influence on agricultural production in the countries that need food the most.

A relatively small percentage of the world's food moves in international trade, and it has been argued that this is not a significant feature of a world food policy. I believe that such a point of view understates the importance of international trade for several reasons. First, the incremental influence of the product traded is greater than is suggested by its percentage of the total. Second, agricultural machinery, fertilizer, and so forth, are also traded. Third, trade provides opportunities for the further development of comparative advantage. This is important for nations that do not have a high comparative advantage in food and fiber production.

4. A major reorientation of research, primarily but not exclusively social and behavioral in nature, is recommended. Such a reorientation need not be costly, compared with research efforts in the natural sciences, but there is a need for clarification of the questions to be investigated.

Much of the research and technical assistance of the social and behavioral sciences is of narrow scope in regard to production and consumption problems in the developing countries. Such efforts are likely to produce fragmentary results and preclude searching examination of the kind of problem mentioned in point 2.

Examples of research issues that I believe should be on the agenda for high-priority examination are the following:

Investigations which will lead to a better understanding of population growth under a variety of conditions. Happily there is a growing body of literature from which generalizations can be drawn (for a survey see Ridker, 1976).

Comparative analyses showing how nations and states have adjusted to particular resource endowments. We need to understand why places such as Sri Lanka and the Indian state of Kerala have significantly outperformed neighboring countries with comparable resources.

Investigations which will lead to a better understanding of the compatibility of the objectives of systems and organizations with those of the larger society. We make a mistake, I believe, in concerning ourselves so much with the output and price performance of large corporations as a measure of competitiveness. They would not be large if they were not reasonably efficient. The real concern comes when large organizations, whether they are public or private, begin to influence the behavior of the consumer or user of their products. For example, as long as the Corps of Engineers or the Bureau of Reclamation could project future requirements and use them as the principal justification for their actions, their future was bright. When it became generally understood that such projections tended to be self-fulfilling, the power of these agencies was significantly reduced. Advertising tends to perform a similar function in the private sector. We need a better understanding of the linkage between the manipulation of wants and the production for the satisfaction of those wants. It is not just consumption stemming from large numbers that is important. The per capita resource consumption of the affluent societies is also of great concern. The problem is not so much that natural resources will be depleted in a physical sense but that the mass transformation of materials places immense pressure on nonmarket resources such as space and the atmosphere.

These investigations will have the following characteristics: (1) They will be concerned with social control mechanisms. It is encouraging that social scientists have made important beginnings in this respect.[4] (2) They will deal with the flow of information between large systems and individual decision makers. Thus, these investigations will have both macro and micro attributes. (3) They will constitute a significant change in perspective, not only for those concerned with food and fiber but for those who work on resources generally. That is, they will tend to concentrate on the determinants of consumption over time. We are entering an era in which more and more questions will be raised about the purposes served by additional consumption and the merits of alternative consumption patterns. We typically refer to this entire area of thought by such nonoperational terms as "life style" or "no growth."[5] Neither of these terms is precise enough to be helpful in this context. However, they do suggest that there may be merit in examining consumption just as searchingly as we have examined supply and technology with consumer sovereignty assumed. Consumption has been neglected largely because our society is supply and technology oriented and our social science theories are better equipped to deal with other issues. Wants or desires—hence, consumption—are largely "givens" rather than variables to be explained in social science investigations. This needs to change.[6]

SUMMARY

The available evidence suggests a growing worldwide need for food and fiber at least for the remainder of this century and probably well beyond that. Estimates show that the social rate of return on public investments in research to increase output of food and fiber has been high compared with other public and private investments, and current policies aimed at increasing such investments appear to be justified.

The utilization of specialized technical knowledge has in the past been associated with large systems in the form of multinational corporations, large governments, and global transportation and communication networks. The interactions among these systems and the autonomous decision makers within them are highly complex and not well understood. It is not surprising that fundamental cause-and-effect relationships are poorly grasped and that social-incentive systems reflect these realities poorly. Because of the mass adoption of technology, the consumption of previously free resources may seriously limit the quality of human life; or the technology that would increase food output may not be adopted, because prices are set by the political rather than the economic system.

The failure to understand and perceive the long-range performance of complex social systems results in government policies excessively oriented toward short-term goals which can be widely perceived and appreciated. Low-cost food and a high rate of employment are examples of such short-term goals.

Our research needs to look beyond increased supply and improved technology. It should be concerned with understanding the linkages between systems, the reconciliation of different perceptions among decision makers in our society, and the factors which influence the ultimate level of consumption. Unless social control mechanisms are devised which reveal to the individual the full social consequences of his or her decisions, our advances in production and technology will be self-defeating.

Although we must eventually have a major work on a par with Adam

Smith, Karl Marx, or J. M. Keynes, we need not wait for such a grand design in order to do much useful work. Whether one is a follower of Popper or Kuhn, the accumulation of refutations (Popper, 1961) or anomalies (Kuhn, 1970) must be examined before new conjectures can be made or a scientific revolution can occur.

Acknowledgment

I am grateful to Kerry Smith, whose comments on an earlier draft greatly improved the organization and rigor of this version.

Notes

1. There is some question whether the renewable-nonrenewable dichotomy is especially useful in resource policy. What may be renewable biologically may not be renewable on economic or social grounds. For a justification of this point of view, see Robert S. Manthy's article, "Scarcity, Renewability and Forest Policy," *Journal of Forestry* (April 1977).

2. Manthy uses methods and data comparable to those of Barnett and Morse. Generally the same comments on methodology and data apply to both.

3. Economists have been highly critical of the Meadows et al. work. Brandow says, "The exercise is an abuse of a method that can be highly useful in appropriate circumstances." Solow writes that the findings are "worthless as science and as guides to public policy." In contrast, Day and Koenig make a number of suggestions for improvement but conclude that if the models stimulate debate on a very important question, their limitations will be forgiven.

4. No attempt is made here to present an exhaustive citation of the literature. The role of social sanctions in an economic system has been the concern of institutionalists such as Veblen, Commons, and Ayres. There is a body of literature which deals with the application of rationalistic models to evaluating institutional alternatives or to deducing behavior in the context of particular rules. The so-called public choice literature is in this vein (Roberts, 1973, p. 387). In addition, there are efforts which are not easily classified. The work of Dunn (1971, 1975) deserves special mention because the incorporation of knowledge acquisition within his research framework permits him to deal explicitly with social learning, which has been neglected or excluded from other investigations. It is en-

couraging to find these different approaches to problems of institutional change. Given the complex nature of the field, there is undoubtedly much merit in a diversity of views.

5. Questions concerning the "life style" of industrialized societies usually focus on the resource-using features of the "affluent" societies. Galbraith's *Affluent Society* (1958) provoked much discussion, but many of his ideas were anticipated by David M. Potter in the early 1950's. In my judgment, the "no growth" concept has tended to divert attention from many legitimate questions about what and how much is consumed (*Daedalus,* Fall 1973). "No growth" advocates are motivated by concern for a legitimate problem. But because this concept is without empirical counterpart in social or biological systems, it rapidly becomes a straw man in the hands of competent analysts.

6. In the context of the report of Study Team 7 of the World Food and Nutrition Study, one can find examples of proposed research which are consistent with the orientation suggested here. The report says that such research can be undertaken without additional disciplinary advances. Of course, increased disciplinary knowledge would undoubtedly result (National Academy of Sciences, 1977b).

References

Barnett, Harold J., and Chandler Morse. 1963. *Scarcity and growth: The economics of natural resource availability*. Publ. for Resources for the Future by the Johns Hopkins University Press, Baltimore, Md.

Brandow, G. E. 1974. Review of Dennis L. Meadows and Donella H. Meadows, eds. *Toward global equilibrium: Collected papers*. In *Am. J. Agr. Econ*. 56(1):193–194.

Bredahl, Maury, and Willis Peterson. 1976. The productivity and allocation of research: U.S. agricultural experiment stations. *Am. J. Agr. Econ*. 58:684–692.

Committee on Agricultural Production Efficiency. 1975. *Agricultural production efficiency*. Board on Agriculture and Renewable Resources, Commission on Natural Resources, National Research Council, National Academy of Sciences, Washington, D.C.

Daedalus. The no-growth society. Fall 1973 (vol. 102, no. 4).

Day, Richard H., and Evan F. Koenig. 1974. Review of Dennis L. Meadows and Donella H. Meadows, eds. *Toward global equilibrium: Collected papers*. In *Am. J. Agr. Econ*. 56(1):194–196.

———. 1975. On some models of world cataclysm. *Land Economics* 51(1):1–20.

Dunn, Edgar S., Jr. 1971. *Economic and social development: A process of social learning*. Johns Hopkins University Press, Baltimore, Md.

———. 1975. Heilbroner's historicism versus evolutionary possibilities. *Zygon* 10(3):272–298.

Galbraith, John Kenneth. 1958. *The affluent society*. Houghton Mifflin, Boston.

Griliches, Zvi. 1958. Research costs and social returns: Hybrid corn and related innovations. *J. Pol. Econ.* 66(5):419–431.

———. 1964. Research expenditures, education, and the aggregate agricultural production function. *Am. Econ. Rev.* 54:961–974.

Hayami, Yujiro, and V. W. Ruttan. 1970. Factor prices and technical change in agricultural development: The United States and Japan, 1880–1960. *J. Pol. Econ.* 78(5):1115–1141.

Hitch, Charles J. 1976. Taking the blinders off. Address to the Brookings Institution Alumni Round Table, December 1976.

Kuhn, Thomas S. 1970. *The structure of scientific revolutions*. 2nd ed. University of Chicago Press, Chicago, Ill.

Manthy, Robert S. 1977. Scarcity, renewability and forest policy. *J. Forestry* 75(4):201–205.

Meadows, Donella H., Dennis L. Meadows, Jorgen Randers, and William W. Behrens III. 1972. *The limits to growth*. Universe Books for Potomac Associates, New York.

National Academy of Sciences. 1977a. *World Food and Nutrition Study: The potential contribution of research*. Washington, D.C.

———. 1977b. *Supporting papers: World Food and Nutrition Study*. Vol. 3. Report of Study Team 7. Washington, D.C.

———. 1977c. *Supporting papers: World Food and Nutrition Study*. Vol. 4. Washington, D.C.

Popper, Karl. 1961. *The logic of scientific discovery*. 2nd printing. Basic Books, New York.

Potter, David. M. 1954. *People of plenty: Economic abundance and the American character*. University of Chicago Press, Chicago, Ill.

Ridker, Ronald G., ed. 1976. *Population and development: The search for selective interventions*. Publ. for Resources for the Future by the Johns Hopkins University Press, Baltimore, Md.

Roberts, Blaine. 1973. An extension of optimality criteria: An axiomatic approach to institutional choice. *J. Pol. Econ.* 81(2):386–400.

Solow, Robert M. 1973. Is the end of the world at hand? *Challenge* 16:39–50.

U.S. Department of Agriculture, Forest Service. 1976. *The nation's renewable resources: An assessment, 1975*. U.S. Government Printing Office, Washington, D.C.

University of California. 1974. *A hungry world: The challenge to agriculture*. Division of Agricultural Sciences, University of California, Berkeley.

2 Land resources and quality of life

Wynne Thorne and Alison C. Thorne

The adequacy of U.S. soil resources to provide ample food and fiber for the citizens of this country with a substantial surplus for export has been reported in numerous publications in recent years (U.S. President's Science Advisory Committee, 1967). Less often emphasized in these reports are the "ifs," "buts," and "perhaps." Some of these limitations and the dynamic changes in the uses of our land are brought out in a recent assessment of U.S. land resources and land uses (Soil Conservation Service, 1976). These data, compiled in 1975, when compared with similar information based on observations made in 1967 (Conservation Needs Inventory Committee of USDA, 1971), reveal vital information about how we are managing our land resources.

In the 1975 survey, Soil Conservation Service specialists examined the use and quality of land at 41,000 randomly selected sites in 506 countries in 50 states, Puerto Rico, and the Virgin Islands. For the 50 states they found 162 million hectares of cropland in 1975 compared with 177 million hectares reported for 1967 (Table 1).

The 1975 survey estimated that there are 45 million hectares not now in crops that could be shifted to cropland if needed (Table 2). Of this, 9.7 million hectares of "prime farmland" could be placed under cultivation without any special development programs and should be capable of pro-

TABLE 1. Condition of U.S. cropland (1967 and 1975).

Problem	Land Class	Cropland (1,000 hectares) 1967	1975
No Problem	1	14,681	13,512
Erosion	2E	36,555	35,449
	3E	32,500	28,471
	4E	14,556	12,017
	5 & 6E	4,904	3,559
	Subtotal	88,515	79,495
Excess Water	2W	23,234	24,329
	3W	13,715	12,698
	4W	2,545	1,606
	5 & 6W	1,096	833
	Subtotal	40,590	39,466
Soil Condition	2S	7,733	8,276
	3S	6,041	4,636
	4S	2,660	2,389
	5 & 6S	1,342	1,430
	Subtotal	17,776	16,731
Climate	2C	8,260	7,976
	3C	5,093	3,910
	4C	367	135
	5 & 6C	106	146
	Subtotal	13,826	12,167
	7 & 8	1,700	670
	TOTAL	177,088	162,041

Source: Soil Conservation Service (1976 and 1967).

ducing high yields. An additional 22 million hectares with "high potential" for crop production could be placed in crops with special practices for erosion control and removal of excess water. The remaining 13.3 million hectares have a "medium potential" for conversion to crops and would have more serious erosion and water disposal problems.

Areas of potential cropland vary substantially in the different regions of the country. They are distributed 17 percent in the Southeast, 16 percent in the Northern Plains Region, 13.4 percent in the corn belt, 12.6 percent in the Mountain Region, and about 10 percent each in the Appalachian, Delta, and Southern Plains regions. In the Southeast the potential cropland is 7,624,000 hectares compared with 6,688,000 hectares now cropped.

Of added interest, of the uncropped land having high and medium potential for cropping, about 15 percent, or 6.6 million hectares, is now covered with high-density forest. Serious consideration should be given to alternative needs for cropland versus timber before this high-density forest is cleared and the land used for crop production.

This 1975 estimate of 45 million hectares of reserve potential cropland is far below the 1967 estimate of 108 million reserve hectares. Consequently, our present 162 million hectares of cropland represents nearly four-fifths of the total area available and suitable for cropping in this country.

When we consider that our harvested cropland has increased more than 16 million hectares in the last four years, our reserve of arable land is be-

TABLE 2. Condition of potential cropland compared with 1975 cropland (in 1,000 hectares).

Problem	Land Class	Potential for Cropland: High	Potential for Cropland: Medium	1975 Cropland
No Problem	1	2,060	139	13,512
Erosion	2E	8,089	1,518	35,449
	3E	6,977	3,482	28,471
	4E	2,374	1,775	12,017
	5 & 6E	1,296	969	3,559
	Subtotal	18,735	7,744	79,495
Excess Water	2W	3,863	837	24,329
	3W	2,437	1,042	12,698
	4W	678	636	1,606
	5 & 6W	209	868	833
	Subtotal	7,187	3,383	39,466
Soil Condition	2S	857	108	8,276
	3S	954	509	4,636
	4S	341	318	2,389
	5 & 6S	180	439	1,430
	Subtotal	2,332	1,374	16,731
Climate	2C	775	295	7,976
	3C	492	108	3,910
	4C	—	137	135
	5 & 6C	93	105	146
	Subtotal	1,360	645	12,167
	7 & 8	—	—	670
	TOTAL	31,674	13,285	162,041

TABLE 3. Changes in land use: 1967 compared with 1975 (in estimated hectares x 1,000).

		Land Use in 1975					
Land Use in 1967	1967 Total	Cropland	Pasture and Range	Forest	Other Land	Urban	Water
Cropland	174,520	142,311	21,401	3,345	5,252	1,961	250
Pasture and Range	205,122	12,913	179,017	5,705	5,738	1,299	450
Forest	179,909	4,463	25,281	141,109	6,395	1,790	871
Other Land	23,177	2,360	5,332	1,783	10,876	1,682	1,144
1975 Total	582,726	162,045	231,031	151,942	28,261	6,732	2,715

Source: Soil Conservation Service (1976).

coming precariously low. To add to our concern, the 1975 data reveal that each year we are losing more than 1 million hectares of farmland to urban buildup and related uses (Davis, 1976). Furthermore, nearly half of our current 162 million hectares of cropland requires special management to restrict erosion. Also, almost 60 percent of the 45 million hectares of potential cropland has such erosion hazards.

The SCS data reveal a dynamic series of changes in land use (Table 3). Of the 175 million hectares classed as cropland in 1967, 142 million are still cropped, 21 million hectares have been converted to pasture and range, 3.3 million hectares have been changed to forest, 5.3 million have been shifted to other uses, 2 million to urban buildup, and 0.25 million to water storage.

There are no comparable data on world land resources. Most world data are based on the 1967 "World Food Problem" report and on FAO estimates, both of which lack recent detailed on-site observations. All such estimates agree, however, that the large undeveloped arable land resources are in Latin America, Africa, and Australia. Although there are substantial areas of potential croplands, they are generally located away from centers of population. In view of land evaluation statistics in the United States, we should consider estimates of large world reserves of potential croplands as very optimistic. And because of their location and the cost of development, they will be utilized very slowly.

The University of California study of world food resources (1974) provides estimates of the annual rate at which present cropland is being expanded between 1970 and 1985. On a worldwide basis this is 0.68 percent per year and varies from an annual 0.44 percent decrease in cultivated area for Europe to a 6.7 percent increase for Oceania (mostly Australia). Latin America is estimated to be expanding by 1.6 percent per year, Africa by 0.9 percent, Asia by 0.25 percent, North America by 0.44 percent, and the USSR by 0.54 percent. If such estimates are reasonably accurate, increased food supplies, with an estimated need for an annual increase of 3 percent, must come primarily from improved production on land already being cropped. This is especially true for Europe, Asia, North America, and the USSR.

The impact of energy needs on land use cannot be clearly foreseen. One important consideration is the possible increased use of wood materials as a source of energy and of industrial materials. In anticipation of such needs the National Research Council Committee on Renewable Resources for Industrial Materials (1976) recommends that "the USDA should immediately develop production-scale demonstration tree farms that use the best available technology for achievement of maximum materials productivity." Such cultivated and frequently harvested tree farms would compete with food and other fiber crops for arable land. While this proposed use of our better soils is still tentative, the potential national importance of such a program is too great not to be a factor in land-use planning. Other nations with limited oil resources will be faced with similar alternatives for producing wood as one substitute.

CLIMATE

Numerous studies have emphasized the close dependence of crop yields on weather (Committee on Climate, 1976). Evidence is seen in the drastic decline in crop production in India associated with a weak monsoon in 1972 and 1973. Almost simultaneously there was a severe drought during 1972–74 in the Sahel region of Africa, and during the same period the USSR had one of its periodic droughts and chose to purchase grains on the world market rather than depend on low internal supplies. In consequence, reserve supplies of grains were largely exhausted and there was near chaos in the world wheat market during 1972 and 1973.

Public concern about weather and future food supplies has been heightened by publicity of reports by Bryson (1974) and others that in the past 70 years the earth, or at least much of the Northern Hemisphere, has had the most favorable weather for crop production that has occurred in the last thousand years. Such publicity has created an impression that the earth will experience less favorable weather in the future, and this impression has been reinforced by reports of a cooling trend attributed to increased particles in the atmosphere.

The Committee on Climate and Weather Fluctuations and Agricultural Production, in their 1976 report, made several important points on the problem of weather and food supply. Agriculture can adjust to long-term trends in climate, but it is very difficult to respond to short-term fluctuations. Also, the predictive accuracy for climatic conditions a few weeks or months in the future is very low, and a cooling trend in one area is usually associated with a warming period in another. We seem to be left in a state of uncertainty about future weather, but this is the usual relationship between man and climate.

Weather fluctuations create problems. India, for example, has had repeated fluctuations in precipitation. Since 1974, however, the climate and crop yields have been so good that in 1975–76 a major problem was adequate storage facilities for the 15 million tons of surplus cereal grains. Meanwhile, the population has been increasing by 13 million per year. Another drought cycle will require substantially greater food reserves than ever before.

LONG-TERM PRODUCTIVITY OF SOILS

The ability of soils to support high crop yields under intensive farming has been a major public concern for centuries. Various theories have even explained the rise and fall of civilizations on the basis of changes in the fertility of their soils.

There is evidence of large areas of salty abandoned lands in the Tigris and Euphrates river basins where civilizations such as Sumer and Babylon flourished and disappeared. Drainage was unknown. Consequently, salts in irrigation waters accumulated in soils in harmful concentrations until eventually the fields were abandoned. Drainage tests in the vicinity of the site of ancient Babylon and near the present city of Kut, to the south, have shown that in recent times good crop yields were possible after installation of drains and leaching to remove the salts. Barley in unreclaimed soils yielded about 0.35 ton per hectare, and on nearby drained and leached land yields of about 2.7 tons per hectare have been obtained.

The Chinese civilization is as ancient as the Tigris-Euphrates civilization. In 1908, Dr. F. H. King, soil scientist of the University of Wisconsin, traveled extensively in Japan, Korea, and China to see how four thousand years of continuous farming had affected variety, yield, and quality of crops produced. He observed the careful collection of animal and human wastes and the composting of crop residues and plant materials gathered from noncropped areas. These organic materials and sediments from canals and rivers were added periodically to the fields. This careful conservation of wastes and plant materials resulted in fertile fields, even though they had been intensively farmed for thousands of years. He recorded average yields of wheat of 1.6 to 2.0 tons per hectare and of 2.1 tons per hectare of paddy rice. Under unusually favorable conditions, a yield of 7.8 tons per hectare of wheat was reported.

The common experience of farmers has been that, except for desert soils that have an initial low content of organic matter, when new lands are placed under cultivation without the use of fertilizers, there is a gradual but pronounced decline in crop yields. Colonists in America repeatedly abandoned so-called worn-out soil to begin anew on virgin land. However, we have direct evidence from long-term field experiments that well-managed soil does not wear out and that depleted soil can be rapidly restored to high productivity by treatment with fertilizers.

In 1843 and succeeding years, Sir John Bennet Lawes and Sir Henry Gilbert initiated a series of systematic field experiments in Bedfordshire, England, to determine how soils could be maintained in a fertile state (Rothamsted Experimental Station, 1974). Plots of soil were established and separately managed. Some plots were left untreated while others received lime, farmyard manure, phosphate, potassium salts, and nitrogen-containing materials—separately and in combinations. In different series of experiments, wheat and barley were grown continuously and in rotations including turnips, beans, mangels, clover, and fallow. These experiments, with some carefully planned modifications, have been continued to the present day. We thus have 134 years of evidence.

The Broadbalk experiments with wheat, at the Rothamsted Station, illustrate the general trends. The plots that have received neither manure nor fertilizer since 1843, and were planted continuously in wheat, currently yield about 2.1 tons per hectare. Those plots of wheat planted in rotation following potatoes and beans, without fertilizer, now yield about 2.8 tons per hectare. Some parts of the plots producing wheat continuously without fertilizer treatment, but which have in recent years been treated with complete fertilizers, have increased in production to 5 and 6 tons per hectare, or nearly three times that of the plots receiving no treatment. Also, the plots that have continued over the years with the modest fertilizer or farmyard manure treatments, started in 1843, have increased in yield by almost 100 percent in recent years compared with the average of the first 50 years. This improvement has been found to reflect improved crop varieties, improved tillage practices, and the control of pests and weeds. It is also observed that in recent years, with improved tillage and weed-control practices, plots treated with farmyard manure now yield somewhat more than those receiving equivalent amounts of nutrients in inorganic fertilizers.

Comparable results were obtained with barley at the Rothamsted Experimental Station in the Hoosefield experiments from 1852 to the present. After more than 120 years, continuous planting of barley on untreated plots produced several symptoms of nutrient deficiency and yields of about 1 ton per hectare. Barley grown on plots treated with farm manure plus 96 or 144 kilograms per hectare of nitrogen as nitrochalk, or on plots receiving potassium, phosphorus, sodium, and magnesium salts plus 96 or 144 kilograms per hectare of nitrogen, yielded 5 to 5.6 tons of grain per hectare.

Similar trends have been observed in long-term field experiments in Illinois and Ohio. Best known are the Morrow plots established in 1876 by Manley Miles at the University of Illinois (Welch et al., 1976). Originally there were ten plots, each of one-half acre. Only three plots have been continued to the present. One plot has grown corn continuously, one a rotation of corn and oats (soybeans have replaced corn since 1967), and the third has corn, oats, and clover. Fertilizer treatments have been superimposed on the rotation plots.

On untreated soil, corn grown continuously has averaged 2.83 tons per hectare since 1955. In the corn-oats-clover rotation, corn yields averaged 5.1 tons per hectare. Subplots of corn grown continuously, which have received lime plus nitrogen, phosphorus, and potassium fertilizers since 1955, have current yields of about 8.4 tons per hectare. Subplots receiving high rates of nitrogen since 1967 have attained yields of 10 tons per hectare. In a corn-soybean rotation, since 1967, similar fertilizer treatments in combination with high rates of nitrogen have resulted in yields of 11.5 tons per hectare.

The high yields obtained by adding adequate nutrients to low fertility plots producing corn continuously in Illinois, or wheat or barley in England, are convincing evidence that soils can continue to produce high yields if fortified with plant nutrients that may have become deficient.

EROSION

The durability of soils is based on their remaining in place. Much of the long-farmed soil of China, Egypt, and Mesopotamia is in nearly flat deltas or flood plains that have little erosion and in many instances have periodically received additions of silt from overflowing rivers. Most cropped lands of the United States are upland soils with some slope hazard and therefore subject to erosion. The 1975 SCS survey indicated that 79.5 million hectares out of 162 million hectares of croplands have a moderate to severe erosion hazard (Table 1). Furthermore, 60 percent of the 45 million hectares of uncropped but potentially arable land in the United States has similar erosion hazards (Table 2).

Without doubt, erosion is the most important factor in limiting long-term soil productivity. Pimentel et al. (1976) have estimated that for soils with 30 centimeters of topsoil, the loss of 2.5 centimeters of topsoil results in a decrease in corn yields by 0.27 ton per hectare per year. This amounts to an average loss of 2.5 centimeters of topsoil in 14 years. Since the rate of soil formation is estimated to average 2.5 centimeters in 30 years, the topsoil of the United States is being lost at about twice the rate it is being formed.

Calculations by Wadleigh (1968), based on soil-erosion data, suggest that more than 45 million tons of the major fertilizer elements—nitrogen, phosphorus, and potassium—are lost annually in eroded soil. The cost of replacing the nitrogen, phosphorus, and one-fourth of the potassium would be about $7 billion per year. Overall, Pimentel et al. (1976) estimate that soil-erosion losses have reduced the production potential of U.S. cropland by 10 to 15 percent.

A study in northern Illinois illustrates a major reason why more effective measures for erosion control are not more widely practiced. Swanson and Harshbarger (1964) estimate that on land with a 4 percent slope, planting corn on the contour in a five-year rotation of corn-corn-oats-meadow-meadow would reduce soil losses to 4.9 tons per hectare. But the practice would cost about $96 per hectare over a 20-year period. In consequence, farmers generally choose immediate net income rather than investing in soil conservation measures.

New soil-management and weed-control technologies, such as reduced or no-tillage systems, are becoming widespread in the United States and Europe (Triplett and Van Doren, 1977). For carefully managed fields, production costs have been cut and erosion has been reduced almost to zero in trials with corn planted in herbicide-killed meadow sod on up to 20 percent slopes. The U.S. Department of Agriculture has estimated that by 2010 more than 90 percent of the area in crops will be grown with reduced tillage systems, and that some form of no-tillage farming will be used on half of the cropped land. Although limited supplies of fossil fuels will increase the costs of herbicides, the small quantities applied per hectare will still make their use in such systems highly advantageous.

Soil erosion in developing countries has been estimated to be nearly twice as severe as in this country (Ingraham, 1975). My own experience in India indicates that erosion is a severe problem in the approaches to the Himalaya Mountains and on the Deccan Plateau that stretches southward from near Delhi. The principal conservation measure has been the construction of thousands of kilometers of bunds and narrow-base terraces on approximately the contour. Because of frequently poor design and inadequate maintenance, the bunds often contribute more to erosion than to its control. Precipitation runoff follows bund channels until a low spot or a break in the bund allows the water to go downhill, creating a gully. Since so much of the soil in these large areas is very shallow over partly weathered granite or other rock materials, erosion is a critical factor.

Clearly, erosion is a crucial worldwide problem. It is a hazard to future food production and is not receiving adequate attention.

TECHNOLOGY AND COSTS

Those who estimate the extent of future food supply commonly assume that new technologies will evolve to continue the recent trend of increasing average yields of food crops. This increase has been near 3 percent per year for corn and about 2 percent per year for other cereal crops. Increases for soybeans have been near 1 percent per year, and most other legumes have not displayed definite trends toward increased yields.

The major technological improvements in the past have included improved cultivars for all major crops, improved fertilizers used in larger quantities, the use of herbicides to control weeds, and improved pesticides and management practices to control insects and diseases. Also, there has been a significant and continuous increase in the area of land irrigated. Improved power equipment and tillage, seeding, and harvesting equipment have also made possible improved and better-timed farming operations.

Application of present technology, combined with favorable moisture and temperature conditions, is capable of producing yields of most major crops two to three times the present national averages (Wittwer, 1974). Since optimum climatic conditions do not generally occur, it is difficult to estimate how much improved farm practices are likely to increase average yields for the nation. Furthermore, costs of production may limit use of some technologies.

Certain disturbing signs indicate a need for caution in such optimistic predictions. Plotting the index of average crop yields for the United States against the quantities of fertilizer being applied (Committee on Agricultural Production Efficiency, 1975) reveals a leveling trend since about 1965. Overall, we are apparently getting decreasing returns from higher increments of fertilizer applied. Furthermore, the energy crisis and escalating costs for fuel, fertilizer, pesticides, and other inputs suggest that lower levels of technology and higher costs of food may be the significant trends in the future. Fertilizer usage in the United States declined about 9 percent in 1975 below the 1974 level.

Clearly the ability of the United States to produce food for its people is not an issue for the current century. Out of the 136 million hectares of harvested crops, 40 million produced goods for export in 1975. The number of harvested hectares used in domestic consumption per capita during the past 30 years has been reduced nearly half, from 0.8 to 0.4 hectare per capita (Economic Research Service, 1976).

QUALITY OF LIFE

Soils and Human Health

In the United States there seems to be more public concern about the nutritional value of food than the possible future inadequacies in world supply. Apparently because of the association of dark productive soils, high in organic matter, with good yields of attractive crops, there are many individuals and groups who believe that the treatment of soil with organic matter results in food products whose nutritional value is superior to those grown in soil treated with chemical fertilizers.

Many studies have been conducted on the relation between soil management and crop production practices and the nutritional value and yields of food crops. To agricultural scientists the answers seem clear and straightforward. But lay groups who are impressed by literature on organic gardening and "natural foods" remain unconvinced.

Farmyard manure, compost, or plant residues added to soil are decomposed by the large and diverse population of soil microorganisms. Intermediate products or decomposition help hold soil particles together and impart to the soil a good tilth that improves soil aeration and tillability and helps the soil to resist erosion. However, plants are unable to absorb particles of organic matter or even appreciable amounts of the smaller organic molecules produced in the decomposition process. The eventual end products of organic matter decomposition are carbon dioxide, water, ammonia, and various mineral elements in forms closely similar to those provided in mineral fertilizers.

In general, then, it is the quantity of plant nutrients in the soil that is important in stimulating the growth of plants and keeping them in a healthy condition. The source of the nutrients, whether from organic or inorganic treatment, is of less consequence.

Adding inorganic fertilizers to soils of low fertility stimulates increased plant growth. Roots in soil contain almost as much organic matter as the tops. If only the seed or fruits of the crop are harvested, fertilizer treatment results in a substantial increase in the quantity of organic matter added to the soil. This increase may exceed the amount commonly added in farmyard manure or compost.

The helpful role of organic matter in improving soils is unquestioned. But to make it the only source of plant nutrients or to assume that it provides unique nutritional properties to plants, is without foundation and would reduce agricultural productivity if widely adopted.

Experimental evidence (Committee on Nutritional Misinformation, 1976; Allaway, 1975) indicates that selecting crops and breeding them for higher levels of protein, specific vitamins, and other ingredients is the best way to improve their nutritional value. There are a few nutritional deficiency diseases, however, that are associated with low quantities of essential mineral elements in the soil. Zinc, iodine, cobalt, and phosphorus are occasionally deficient in soils, and deficiencies may carry over to animals and human beings. Soil treatments or food additives can correct such problems, and in the case of phosphorus, and often zinc and cobalt, soil treatment also results in substantial increases in crop yields.

A few nutritive substances are affected by environmental factors. Adequate sunlight favors increased ascorbic acid in fruits and leafy vegetables. The thiamine content of plants is also increased by sunlight.

Experimental data, in general, indicate that plants that grow normally and produce satisfactory yields will contain the vitamins characteristic of the species and variety. There is no evidence that food crops treated with organic fertilizers are better quality than those treated with inorganic fertilizers.

Food Consumption Patterns

An increasingly capable agriculture, together with technical improvements in storage, has brought a varied diet to the people of the United States compared with the corn and pork diet of the 1800's (Bell, 1967).

Not too little but too much food has become the problem of some of our people, as is evidenced by the incidence of obesity and heart ailments. Furthermore, we waste a great deal of food. The USDA food plans, which are priced regularly throughout the country, are partly defined in terms of discarded edible food. The low-cost diet allows for one-tenth discard, the moderate plan permits one-sixth discard, and the liberal plan includes one-fourth discard of edible food (U.S. Department of Agriculture, *Agricultural Statistics, 1975*).

Also, the energy costs of getting food to the consumer's table are great. Only one-fourth of the energy used in the U.S. food system goes into farm production; the rest is used to transport, process, preserve, and distribute food, and then the average American shopper drives a two-ton automobile a short distance to a supermarket at least once a week to transport perhaps 30 pounds of food (Brown, 1974).

Americans are fond of meat; per capita consumption in 1974 was 75 kilograms (*Agricultural Statistics, 1975*). Yet livestock production uses up many more resources than do soybeans, also a high-protein food. Heady and Timmons (1975) estimate that if soya food products were substituted for as much as 25 percent of projected beef consumption in the year 2000, as much as 38 million hectares of our cropland base would be available for other uses, even with a population of 300 million people.

The average North American currently uses up to five times as many agricultural resources as the average Indian, Nigerian, or Colombian (Brown, 1974). It is possible that such a disparity, if widely publicized, might lead to some voluntary changes in the consumption habits of the American public.

Life Styles

In 1900 nearly half of the U.S. population lived on farms. Since then there has been a steady decline. From 1959 to 1974 the number living on farms fell from 9.4 to 4.4 percent (*Agricultural Statistics, 1975*). This may be a tribute to the efficiency of agriculture, in sharp contrast to countries such as India where 75 percent of the population lives in rural communities.

But over the years, life has been hard for many U.S. farm families. Throughout the history of the country they have suffered more economic recessions than urban dwellers; and electrification came later to farms than to towns and cities. The rural poor have been very poor indeed, as *The Grapes of Wrath* (Steinbeck, 1939) portrayed in the 1930's, and *The People Left Behind* summarized in the 1960's (U.S. President's National Advisory Commission on Rural Poverty, 1967). In terms of 1969 income, 16.4 percent of families living on farms were below the poverty level, and 15.1 percent of the rural nonfarm families also fell below. This is in contrast to the national urban poverty rate of 9 percent. Half of all black families in rural areas were below the poverty line in 1969 (U.S. Department of Commerce, 1970).

Barkley (1976) insists that the economic role of the family farm in U.S. society must be better understood. The family farm is a national asset because of its ability to absorb adversity and thus to stabilize food supplies, but we do not know how much adversity can be absorbed: "Farm family labor has likely been dreadfully underpriced, thus exaggerating the claims other resources have on the farm's returns."

A surprising number of farm families have other sorts of employment they engage in to keep themselves afloat. This is partly because they give a high priority to living close to the soil. Many other people also show signs of wanting a rural life. Evidence of their nostalgia for the land is seen in the great interest in Currier and Ives prints, the sales of early versions of the Sears, Roebuck catalog, the popularity of the *Foxfire* books and the *Whole Earth Catalog,* and the establishment of rural communes.

The value of work on rural communes is not measured by productivity alone, but also by how this work contributes to group solidarity and personal growth. Cottage industries are often supplementary, and are chosen because they encourage group cohesion. Light farming is preferred because it involves physical labor, a minimum of differentiating skills, a more-or-less immediate product, low initial capital, an abundance of tasks that many people can do together, and an attunement to the rhythms of nature (Kanter, 1973).

Theorists suggest that the commune movement is strongest in times of dislocation. In the United States, communes flourished from 1820 to 1860, and then again in the late 1960's. Over the last few years new rural communes have been formed at a slower rate, although interest in rural living appears to be the same or greater. At present, forms less socially intense are gaining prominence, such as family homesteads, cooperative villages, and neighborhood revitalization (Gardner, 1973).

There is statistical evidence of a move to nonmetropolitan areas. During 1970–73 these areas gained 4.2 percent in population compared with 2.9 percent for metropolitan areas. The attractiveness of small towns has increased, partly because of decentralization of industry and the growth of recreation and retirement activities. Clusters of nonmetropolitan retirement communities are to be found in old cutover regions of the upper Great Lakes, the hill country of central Texas, and the Sierra Nevada foothills of California (Beale, 1975).

It is a curious fact that in the developed countries urban people yearn for the rural life, whereas people in poorer countries yearn for the cities because they hope to find there the jobs they so desperately need. Our current back-to-the-land movement seems to be a characteristic of an affluent society which has decided that "material" things are not as important as "things of the spirit." However, "things of the spirit" are undergirded by scarce resources.

For people who cannot move bodily to rural areas, there is the satisfaction of gardening in town and city. Vegetables are to be found growing everywhere, from flower boxes to plots beside commercial buildings. Publications on gardening, food preservation, and kindred matters are flourishing. The shortage of canning lids during two recent summers is evidence that food preservation is being taken seriously. While some of this activity is motivated by high food costs, some of it is done for pleasure.

At the turn of the century, home gardening was a matter of survival, especially in the coal and steel regions; consequently, the grounds around urban and suburban houses sometimes looked like rural farmyards. A study of 2,500 families living in the principal coal, iron, and steel regions in 1890 suggests that about half of them had livestock, poultry, or vegetable gardens, or all three. Nearly 30 percent purchased no vegetables other than potatoes during the course of the year (U.S. Commissioner of Labor, 1891).

Amenity Services

Suburbanites want clean air, clean water, and open space. People want to see beauty in the form of green strips along the highways. As farmland is converted into residential areas, and shopping districts are built to serve these areas, amenities are threatened. As more people crowd on the land, everyone's freedom becomes restricted, and freedom is an amenity.

It is obvious that land produces commodities, such as food and forest products, and it provides space for buildings to stand on, but it also produces amenity services to answer recreational and aesthetic needs. René Dubos (1967) has said: "Just as important for maintaining *human* life is an environment in which it is possible to satisfy the longing for quiet, privacy, independence, initiative, and open space. These are not frills or luxuries but constitute real biological necessities. They will be in short supply long before there is a critical shortage of the sources of energy and materials that keep the human machine going and industry expanding."

The U.S. Bureau of Land Management and the U.S. Forest Service administer 263 million hectares of land. And there remain as many as 20 million hectares of roadless woodland areas not under statutory protection. Krutilla and Fisher (1975) point out that to determine the wisest use for such lands is more than a matter of cost-benefit analysis. It means considering the recreational, aesthetic, and other satisfactions associated with these lands. It means retaining an option which would otherwise be foreclosed. There is the purely scientific interest, and the importance of natural biota which represent our reservoir of genetic information; there are botanical specimens and medicine. But there is also a psychic basis, in that many persons obtain satisfaction from the knowledge that part of the wilderness of North America still exists, even if they do not choose to go there themselves. These are amenity services, in contrast to commodity services, and their value seems to be rising relative to commodity resources.

Ingraham (1976), by spelling out the actual situation on the Running Creek watershed on the Front Range of Colorado, shows the need for advanced thinking before particular crises in land use arise. She notes that this long, narrow watershed contains elements of virtually all the problems facing the Interior West, as well as many other regions of the nation. These problems include "urban pressures on agricultural land, recoverable energy resources, proximity to a major metropolitan area, potential as a corridor for transportation and transmission, serious water constraints, increasing potential for environmental deterioration, unequal distribution of goods and services, jurisdictional conflicts for land-use planning, heavy dependence on weather, and marginally productive lands."

Considering that the world is now facing shortages of the four basic agricultural resources of land, water, energy, and fertilizer, present and future trade-offs will be critical. Yet the ultimate problem is not so much absolute scarcity in land supply as it is how land and related resources will be allocated among agricultural and other alternatives. The question is also how to maintain a reserve of land resources so that options will be available in the years to come.

References

Allaway, W. H. 1975. The effect of soils and fertilizers on human and animal nutrition. U.S. Dep. Agr. Inf. Bull. 378.

Barkley, Paul W. 1976. A contemporary political economy of family farming. Am. Agr. Econ. Assoc. meetings, August 1976. Pennsylvania State University, University Park, Pa.

Beale, Calvin. 1975. The revival of population growth in non-metropolitan America. U.S. Dep. Agr. Econ. Res. Serv. No. 605.

Bell, Carolyn Shaw. 1967. *Consumer choice in the American economy*. Random House, New York.

Brown, Lester R. 1974. *By bread alone*. Praeger Publishers, New York.

Bryson, Reid A. 1974. A perspective on climatic change. *Science* 184:753–760.

Committee on Agricultural Production Efficiency. 1975. *Agricultural production efficiency*. Board on Agriculture and Renewable Resources, Commission on Natural Resources, National Research Council, National Academy of Sciences, Washington, D.C.

Committee on Climate and Weather Fluctuations and Agricultural Production. 1976. *Climate and man*. Board on Agriculture and Renewable Resources, Commission on Natural Resources, National Research Council, National Academy of Sciences, Washington, D.C.

Committee on Nutritional Misinformation. 1976. Soil fertility and the nutritive value of crops. *Nutrition Reviews* 34(10):316–317.

Committee on Renewable Resources for Industrial Materials. 1976. *Renewable resources for industrial materials*. Board on Agriculture and Renewable Resources, Commission on Natural Resources, National Research Council, National Academy of Sciences, Washington, D.C.

Conservation Needs Inventory Committee of U.S. Department of Agriculture. 1971. National inventory of soil and water conservation needs, 1967. U.S. Dep. Agr. Stat. Bull. 461.

Davis, R. M. 1976. What is left of our cropland "frontier"? *Soil Conservation* 42(1):2.

Dubos, René. 1967. Man adapting: His limitations and potentialities. Pages 11–25 in William R. Ewald, Jr., ed., *Environment for man*. Indiana University Press, Bloomington, Ind.

Economic Research Service. 1976. Changes in farm production and efficiency. U.S. Dep. Agr. Stat. Bull. 561.

Gardner, Hugh. 1973. Crises and politics in rural communes. Pages 150–166 in Rosabeth Moss Kanter, ed., *Communes: Creating and managing the collective life*. Harper and Row, New York.

Heady, Earl O., and John F. Timmons. 1975. U.S. land needs for meeting food and fiber demands. *J. Soil and Water Conserv*. 30(1):15–22.

Ingraham, Elizabeth Wright. 1976. Lead time for assessing land use: A case study. *Science* 194:17–22.

———. 1975. *Query into the quarter century*. Wright-Ingraham Institute, Colorado Springs, Colo.

Kanter, Rosabeth Moss. 1973. Work and property, introduction. Pages 223–230 in Rosabeth Moss Kanter, ed., *Communes: Creating and managing the collective life*. Harper and Row, New York.

King, Franklin Hiram. n.d. *Farmers of forty centuries*. Harcourt, Brace and Company, New York.

Krutilla, John V., and Anthony C. Fisher. 1975. *The economics of natural environments*. Publ. for Resources for the Future by the Johns Hopkins University Press, Baltimore, Md.

Pimentel, D., E. C. Terhune, R. Dyson-Hudson, S. Rochereau, R. Samis, E. A. Smith, D. Denman, D. Reifschneider, and M. Shepard. 1976. Land degradation: Effects on food and energy resources. *Science* 194:149–155.

Rothamsted Experimental Station. 1974. *Rothamsted Experimental Station guide*. Harpenden, England.

Soil Conservation Service. 1976. Potential cropland survey. Preliminary report. U.S. Department of Agriculture, Soil Conservation Service, Washington, D.C.

Steinbeck, John. 1939. *The grapes of wrath*. Viking, New York.

Swanson, E. R., and C. E. Harshbarger. 1964. An economic analysis of effects of soil loss on crop yields. *J. Soil and Water Conserv.* 19(5):183–186.

Triplett, G. B., Jr., and D. M. Van Doren, Jr. 1977. Agriculture without tillage. *Scientific American* 236(1):28–33.

U.S. Commissioner of Labor. 1891. *Sixth annual report: Cost of production*. Cited in Robert W. Smuts. *Women and work in America*. 1959. Columbia University Press, New York.

U.S. Department of Agriculture. *Agricultural statistics, 1975*. U.S. Government Printing Office, Washington, D.C.

U.S. Department of Commerce, Bureau of the Census. *1970 census of population*. U.S. Government Printing Office, Washington, D.C.

U.S. President's National Advisory Commission on Rural Poverty. 1967. *The people left behind*. Washington, D.C.

U.S. President's Science Advisory Committee. 1967. *The world food problem*. Vol. 2, chap. 7. Washington, D.C.

University of California. 1974. *A hungry world: The challenge to agriculture*. Division of Agricultural Sciences, University of California, Berkeley.

Wadleigh, C. H. 1968. Wastes in relation to agriculture and forestry. U.S. Dep. Agr. Misc. Publ. 1065.

Welch, L. F., S. W. Melsted, and M. G. Oldham. 1976. Lessons from the Morrow plots. *Illinois Research* 18(1):3–4.

Wittwer, S. H. 1974. Maximum production capacity of food crops. *BioScience* 24(4):216–224.

3 Assessment of technology in food production

Sylvan H. Wittwer

Food is a renewable resource. Plants provide directly, or indirectly, up to 95 percent of the world's food supply. Chief among the major food crops, in approximate order of importance, are the cereal grains (rice, wheat, maize, sorghum, millet, barley, oats, and rye), the seed legumes (field beans, peanuts, chick peas, pigeon peas, soybeans, cowpeas, mung beans, and broad beans), the root and tuber crops (potatoes, sweet potatoes, cassava, and yams), the sugar crops (sugarcane and sugar beets), and the tropical crops (bananas and coconuts). Grapes, tomatoes, and oranges are prominent in the diets of many people. Rice, wheat, maize (corn), and potatoes contribute as much food tonnage to the world total as all other crops combined. A great variety of other fruits and vegetables are secondary staple crops that seldom get into economic surveys. Processed and fresh, they add essential dietary nutrients and personal enrichment and pleasure to eating. Hundreds of millions of people depend primarily on what is produced in gardens or small holdings. The consumers are the producers. The surplus is sold in "peoples markets" in the nearby cities or villages. Several crops a year may be grown.

Cereal grains constitute the most important food group on earth. They provide 60 percent of the calories and 50 percent of the protein consumed by man. Seed legumes furnish 20 percent of the protein.

The number of livestock in the world is double the human population. Domestic animals produce meat, milk, and eggs from nutrients derived from crops, forages, and by-products that have less value elsewhere. Food resources produced by animals fall into two categories: those derived from ruminants (dairy and beef cattle, water buffalo, sheep, goats, camels, llamas, alpacas), and those derived from monogastric creatures (swine, poultry, rabbits, guinea pigs). Ruminants provide foods of the highest quality from range and grassland, forages, plant by-products, crop residues, and browse. Swine and poultry are providers of food derived from consumption of many products not edible by man. Livestock in agriculturally developing nations serve as an important source of food, dietary improvement, income, power, fuel, and fertilizer. They are also a living food reservoir, which exceeds the present worldwide commercial grain reserves (40 days compared with 27) and is not concentrated in a few nations with surpluses. Globally, about 25 percent of man's protein requirements and 10 percent of the calories and significant amounts of calcium and phosphorus are provided by livestock and poultry. Domesticated livestock (including poultry) in the United States produce two-thirds of the protein, one-half of the fat, one-third of the energy, four-fifths of the calcium, and two-thirds of the phosphorus consumed by man. The flesh of aquatic animals provides 5 percent of man's protein needs. Fish are by far the most significant single dietary source of animal protein in the tropics.

TECHNOLOGICAL FRONTIERS IN FOOD PRODUCTION

Attention should be directed toward development of technologies to increase efficiency in photosynthate accumulation and partitioning, greater efficiency of biological nitrogen fixation and use by plants, and the development of new techniques for genetic engineering. These should include an expansion of the techniques for conventional plant breeding as well as *in vitro* techniques for asexual approaches and wide crosses.

Crop yields are dependent more than anything else on the net accumulation of photosynthates in usable organs of the plant, and the availability of adequate supplies of nitrogen. Advances in photosynthetic efficiency and nitrogen fixation, in turn, depend on the integration or recombination of new types of genetic material. Recombinant DNA research will become important, along with the specialized containment facilities needed to do this research. The ultimate perfection of tissue culture techniques will provide a rapid means of screening genetic materials for more efficient methods of accumulating and partitioning photosynthates, utilizing applied fertilizer, and improving biological nitrogen fixation. New technologies for these purposes, and for genetic engineering, developing resistance to environmental stresses and competing biological systems, and greater efficiency in nutrient uptake by plant roots, are nonpolluting, without noise, add to the resources of the earth for all mankind, and are nonpolitical. They are permanent and without limit.

Photosynthesis

The earth's greatest unexploited resource is sunlight. Photosynthesis is the most extraordinary mechanism ever devised. Green plants, as yet, are the primary harvesters of free solar energy and the main producers of food and energy on a renewable basis. Agriculture is basically a solar-energy processing machine. It is the only industry that utilizes today's incident solar radiation, and is man's largest current user of solar energy. Agriculture can also play a unique role in the energy balance of the world.

Harvesting the sun, however, requires the management of land, and often irrigation water, fertilizer, and other proxies. The geometry of agriculture demands that crops be distributed over the land in a layer a few inches thick for maximal use of sunlight. Livestock must also be dispersed for most effective use of grazing, pasture, savannah, and prairie lands. The sun strikes everywhere. It is, therefore, the intent in agriculture to collect and store solar energy as food, feed, and fiber in plant and animal products and to do it with utmost efficiency. Man has specialized in and depends on the culture of sun-loving crops.

There are many research possibilities for the enhancement of photosynthesis. They include (1) identification and control of the mechanisms that regulate and could reduce the wasteful processes of both dark- and light-induced (photo) respiration, and the mechanisms responsible for redistribution of photosynthates which in turn regulate yield and maximize the "Harvest Index," (2) resolution of the hormonal mechanisms that control flowering and leaf senescence, (3) improvements in plant architecture and anatomy, cropping systems, planting designs, and cultural practices for better light reception, and (4) carbon dioxide enrichment of crop atmospheres. The future of the world's food supply and much of the energy may well reside at the door of photosynthesis and subsequent partitioning of the products of photosynthesis into the harvested parts. The relative simplicity of this approach belies its credibility.

Biological Nitrogen Fixation

Nitrogen fixation, the second most important and magnificent process on earth, is the primary source of food protein. Agriculturally important for legumes, and some nonlegumes, it also shows potential for many other crops.

New technology in biological nitrogen fixation should begin with legumes, which offer an efficient system with good chances for a quick payoff. Furthermore, legumes make important contributions to improving yields worldwide, and also to maintaining productive food systems in the tropics and subtropics as well as temperate zones. They provide seed pods for man and forage for livestock, and contribute to soil improvement and ground cover to reduce soil erosion. Some legumes have great undeveloped food potential. For example, the pods and seeds of the desert mesquite could be used for food, and all parts of the winged bean plant of the tropics can be harvested for human consumption.

Legumes have rhizobial associations in root nodules which provide a built-in nitrogen source. Nevertheless, legumes still obtain up to 75 percent of their nitrogen from what is already in the soil. Yet that which is fixed in the nodules is rate-limiting for growth. Production of all legumes could be substantially increased if nitrogen fixation in the nodules could be improved. It has been dramatically demonstrated that photosynthates are a significant limiting factor in nitrogen fixation. A major effort should be directed toward increasing the photosynthetic energy source in legumes. A second approach should be to develop super strains of rhizobia for each of the several species that infect the major food legumes, optimize the efficiency with which nitrogen is fixed per unit of energy conserved, and prolong the capability for nitrogen fixation into the seed-filling period. The alternative is to improve the host plant and its environment by exploring the effects of variety, age, pH, temperature, and soil nutrient levels.

Blue-green algae in tropical soils and rice paddies offer a special opportunity. A blue-green algal relationship has been noted between the minute aquatic fern *Azolla* and the algae *Anabena*. The fern becomes "infected" with algae. The amount of nitrogen fixed under this relationship may exceed 600 kilograms per hectare per year. Unlike the process for legumes, this nitrogen fixation is not limited by available energy, since both the blue-green algae and the tiny fern have built-in energy sources. The science, and art, of *Azolla-Anabena* culture poses a fascinating opportunity for improving yields of rice with little or no additional nitrogen fertilizer (Shanmugam and Valentine, 1976).

Genetic Improvement

Some of the most remarkable increases in food crop productivity have occurred as a result of genetic manipulations. The standard techniques of selection based on phenotypic expression, controlled hybridization, and, more recently, selection for better nutritional qualities have given us super strains of rice, wheat, maize, sorghum, millet, some legumes, and many fruits and vegetables of excellent nutritional quality.

Although standard plant-breeding techniques are likely to predominate in the foreseeable future, there are other frontiers. One is a modification of conventional breeding technology developed by International Agricultural Research Centers. It consists of high-volume crossing, broad use of existing genetic material, and wide regional evaluations of large segregating populations by teams of agriculturists.

Significant advances have also been made in defining techniques for isolating protoplasts (plant cells without walls), their fusion, and subsequent regeneration into new plants. These new cellular approaches to plant breeding could produce new species with greater yields, resistance to stresses and toxins, and improved nutritional quality. Results of studies on the transformation and regeneration of parasexual hybrids from fixed protoplasts could revolutionize agriculture.

The collection, preservation, and utilization of genetic materials of food crops is vital in alleviating genetic, climatic, and chemical vulnerability. Merely broadening the genetic base of major food crop varieties will not prevent epidemics or crop disasters. The natural variation that exists in cultivated and wild species must be preserved. Germplasm is the key.

The preservation of genetic materials of the principal food crops is now proceeding on a systematic and worldwide basis. The process is relatively simple for the cereal grains, the seed legumes, forages, and many vegetables. Those that are vegetatively propagated (potatoes, sweet potatoes, cassava, fruit trees) pose a special problem. There is now an effort to freeze meristems in liquid nitrogen for the preservation of genetic stocks of vegetatively propagated food crops. This will require high technology.

Genetic Improvement in Nutrient Components

Cereals dominate the diets of most people. Progress in raising levels of protein and critically deficient amino acids in cereal grains has been singular. There is no cheaper, better, or quicker way to provide the protein needed by people in most agriculturally developing nations than to improve the cereals they eat. Both the biological value and the level of the protein of maize have been enhanced using the opaque-2-recessive gene; and, more recently, such improvements have been made in maize of normal background. Rice, wheat, and barley selections with higher protein levels have been identified. Triticale, a new synthetic species, with its improved nutritional contributions, great adaptability, and high yields, is now receiving limited commercial acceptance.

Recent evidence indicates that the protein quality of cereals is inversely related to their prolamine content, which varies from a low of 5 to 10 percent for rice and oats, to intermediate levels of 30 to 40 percent for wheat and barley, to a high of 50 to 60 percent for sorghum and maize. Grains such as sorghum and maize are photosynthetically efficient (C_4 plants) under hot-dry conditions, whereas grains such as rice and wheat (C_3 plants) are less efficient. Combining the high productivity of cereal crops with superior protein quality, and C_3 grain characteristics, remains one of the great challenges of plant research.

There is a strong indication that the world's nutritional problem is not one of protein deficiency but of caloric inadequacy. If sufficient calories were provided through conventional cereals and seed legumes, and the biological value of the proteins of these crops was genetically upgraded, there would be no protein problem. Indeed, it has been demonstrated that most of the clinically observed "protein malnutrition" is the secondary consequence of a caloric inadequacy that occurs among people who have sufficient protein in their diets but are unable to assimilate it when caloric uptake is inadequate.

This observation should create new interest in the productive potential of the high-energy root crops—sweet potatoes and cassava. These food crops predominate in much of the developing world. Their promotion might well be predicated on the fact that calorie deficiencies are common for hundreds of millions of people in South America, Southeast Asia, and Equatorial Africa. Cassava should not be developed as a major food protein source but as a source of calories. Here the potential is tremendous. Yields of up to 50 metric tons per hectare have already been achieved in Brazil.

Roots, Mycorrhizae, and Nutrient Uptake

Fertilizer manufacture is the most important industrial input into agricultural productivity. Nitrogen fertilizer inputs exceed $1 billion annually in the United States for maize alone. Yet only 50 percent of the nitrogen and less than 35 percent of the phosphorus and potassium applied as fertilizer in the United States are recovered by crops. The recovery of fertilizer nitrogen in the rice paddies of the tropics is only 25 to 35 percent. The balance is lost to the environment. Denitrification loses nitrogen to the atmosphere. Nitrification encourages losses in the soil from leaching. Food production could be greatly improved if these enormous losses, particularly of nitrogen, in the warm soils of the tropics could be reduced. A potential environmental hazard may also be alleviated (Shapley, 1977). Much of the loss of nitrogen occurs from bacteria-induced nitrification and subsequent denitrification. Nitrification inhibitors, both natural and synthetic, applied with ammonia or urea are effective deterrents to leaching and atmospheric losses of nitrogen. Their use is just beginning on a global scale. Expenditures for fertilizer could be greatly reduced. Nonrenewable resources would be preserved and the food supply greatly improved. It has been estimated that biological denitrification alone depletes the soil of 30 to 50 million tons of nitrogen annually. This is almost equal the amount that is chemically fixed as fertilizer. It is not known what fraction of applied fertilizer undergoes denitrification and is lost to the atmosphere.

An important approach to better recovery of soil-applied fertilizer and also to increased crop productivity on marginal soils is a further microbiological one. In sharp contrast to the action of nitrifying bacteria which greatly reduce the effectiveness of fertilizer nitrogen, other microorganisms, the mycorrhizae (specifically the endomycorrhizae), may bring about large increases in the uptake of phosphorus and other poorly mobile ions. Almost all food crops respond. Mycorrhizae are obligate parasites and can be viewed as fungal extensions of root systems. They can help roots absorb nutrients, and they can stimulate growth and nitrogen fixation by legumes, especially in phosphorus-deficient soils. There are super strains of mycorrhizae, and crops can be inoculated with them. Mycorrhizal fungi have been reported to increase significantly the yields of both legumes and cereal grains. Research on mycorrhizae in particular, and soil microbiology in general, has scarcely begun.

Foliar applications of fertilizer have been declared the most efficient method of fertilizer placement. Technology is lacking in this area. Future yield barriers may well be broken by utilizing the absorptive capacities of leaves, as well as roots, at crucial stages of plant development. Extensive experimental foliar spraying of nutrients was conducted in all major soybean-producing areas of the United States during the summer of 1976. The rising costs of nonrenewable fertilizer should be an added stimulus to further development of this technology.

Climate, Weather, and Food

Decreasing the vulnerability of crops and livestock to adverse climate and weather offers great promise. Technology, resources, and policy are involved. Because weather and climate are still the most determinate factors in food production, we should no longer ignore the potential, through research, for precipitation enhancement and the alleviation of major droughts, and for making our crops more climate proof. Year to year, and season to season, fluctuations in temperature, precipitation, and sunlight are far greater than any long-term climate trends that have thus far been identified. Much can be done to alleviate weather and climatic stresses, including crop diversification, improved cultural practices, pest management, judicious use of land, and greater efficiency in use of water resources.

Pest Management

Crop losses from insects, diseases, and weeds are enormous. For cereal grains they range from a high of 46 percent for rice to a low of 24 percent for wheat. The aggregate is 35 percent of potential production for all cereal crops. It is astonishing that the greatest losses occur in the production of rice, the number one food crop on earth. Insects alone in Southeast Asia annually harvest 41 percent of the rice crop and 48 percent of the sugarcane. Losses from pests include in large part the sections of the green plant already synthesized. A 10 to 20 percent reduction in losses from pests would provide food for hundreds of millions of people. Effective pest-control practices could eliminate 50 percent of the current losses in the food-production cycle.

Toxic chemicals have been the principal means of controlling pests, and crop-protection technology becomes obsolete at an alarming rate. More than a billion pounds of pesticides are used in the world each year (National Academy of Sciences, 1975). Meanwhile, systems of integrated pest management are emerging and can be created for any crop. These systems deal with all aspects of pest control, including natural enemies, resistant varieties, cultural practices, and chemicals. This form of environmental management seeks to obtain the greatest return in food production through the least expenditure of resources, by controlling pests at the lowest cost to man and the environment.

Other new ways are being developed to fight pests using reduced chemical and energy inputs. Allelopathy for weed control is one. This is a process, defined as "mutual harm," in which chemicals produced and released by one plant species inhibit the growth of another. No significant breakthrough with this technology has yet occurred for any major food crop.

Chemicals known as bioregulators, which often duplicate genetic effects, can be expected to have an exciting impact on agricultural productivity. There is new hope and interest in improving agronomic crops by using chemical ripeners that increase sugar yields by as much as 10 percent on sugarcane. Better test methods will open new frontiers for useful application.

Food Losses

It has been estimated that losses of energy and wastes in food processing could be reduced by 35 and 80 percent, respectively. Food losses between harvest and consumption could be reduced by 30 to 50 percent. All this could increase the available food by 10 to 15 percent without more land or increased yields. Annual production of crop and animal wastes in the United States exceeds 800 million tons of dry matter. The United States annually produces 650 thousand tons of seafood waste and 35 million tons of fruit and vegetable processing wastes. Globally, approximately 150 pounds (68 kilograms) of cellulose are produced daily for each of the earth's 4 billion inhabitants. This is equivalent to 1.7×10^{11} tons of dry biomass annually, of which only about 1 percent is currently used for fuel and about 1 percent for food and feed. The vegetation of the earth through its photosynthetic ability produces every year 12 times as much energy as all human beings consume. Economic incentives and better technologies are needed to make the best use of these vast resources in food, feed, fiber, and energy-producing systems.

Forage Production and Range Management

Forage and range production is not a direct food source for human consumption, but it provides a food supply for approximately 2.5 billion ruminant animals useful to man. There are two ruminant animals for every three human beings in the world, with 60 percent in the developing countries. Forages (harvested forages, pastures, rangelands, silages, crop residues) are major world crops in terms of value and area, and they include both grasses and legumes. It is estimated that for every unit of cultivated crop in the world there are two of forage. More land is devoted to forage production than all other crops combined, and more than half of the feed consumed by beef cattle, dairy cattle, sheep, and goats in the United States comes from forages. The tropics provide nearly 1.5 billion hectares of land suitable for forages but of little use for cultivated cropping. The chief and often the only feed source for livestock in developing countries is forages. On a worldwide basis, ruminant-animal production is encouraged to improve nutrition. This improvement depends ultimately on advances in the production and management of forages. Forages may also provide needed ground cover to reduce soil erosion, and leguminous forages fix atmospheric nitrogen for soil improvement, provide high forage yields without added nitrogen, and increase the yields of associated grasses.

Current livestock grazing capacity of U.S. ranges is 213 million AUM's. An AUM (animal unit month) is enough dry forage to maintain a cow and calf of 450 kilograms, or seven mature ewes, for one month. It is technically and economically feasible to increase, and in many cases double, this carrying capacity. In the western half of the United States there are 400 million hectares on which it is feasible to increase productivity by 25 to 50 percent. Maximum increase in productivity could raise the present 213 million AUM's to 266 or 320 million AUM's through the use of improved legumes and grasses, reseeding of ranges where appropriate, incorporation of appropriate legumes into rangeland and pasture, reduction of weeds by herbicides, improvement of range management, and judicious fertilization with nitrogen.

Other research might deal with increasing the digestible protein, designing better forage harvesting systems, improving utilization systems, and making greater use of tree crops and shrubs for livestock feeding, particularly in the tropics.

Forage quality is the chief restriction on ruminant production in the tropics. Improving tropical forages is an inexpensive way to increase animal productivity while improving the land resource base and soil structure. Tremendous productivity is possible in the tropics, with annual yields approaching 60 tons of dry matter per hectare. There are three distinct areas, extending from the Tropic of Cancer to the Tropic of Capricorn, remarkably suited for forage production: the lowlands, the high elevations, and the *cerrada* or savannah.

Small increases in the productivity of forages, grasses, and legumes on rangelands, because of their vastness, would result in significant increases in the production of food. Improved tropical legumes, appropriately introduced into pastures, can fix 25 to 100 kilograms of nitrogen per hectare per year. This would have a pronounced effect on ruminant productivity, through improved quality, nutrition, and total production. The benefits could be achieved in both small- and large-scale farming systems. Improved forage harvesting systems would reduce losses, which now approximate 30 to 40 percent of the digestible nutrients, and could increase efficiency of ruminant production in the northern United States by 25 to 30 percent.

Efforts to breed better forages should result in significant increases in yields. Greater palatability would increase consumption sharply. Improvements in digestibility of dry matter are also likely. The eventual overall impact on food production from improvements in the production and handling of forages and rangelands could be as great as those achieved for wheat, rice, and possibly even maize.

Livestock Productivity

Increased livestock productivity can come from improved production and utilization of feeds, disease control, and genetic improvement. Large ruminant livestock offer the greatest potential (Byerly, 1977).

Three major interrelated research goals must be pursued simultaneously: improved production and utilization of feeds, improved animal health, and genetic improvements. All are equally important. Little progress in better use of feeds can be achieved without research on animal health, genetic improvement, and increased reproductive efficiency. Similarly, emphasis on animal health would be of little value unless accompanied by better feeding. In fact, improvement of the breeding stock may be impossible in some agriculturally developing countries until certain diseases are controlled and certain nutritional disorders alleviated.

Enhancement of productivity through animal health must include control of epidemic and endemic diseases, study of the economic impact of animal diseases, and fundamental research on the biological basis for host resistance to disease and parasite infections. Mortality from disease and parasites has exceeded 50 million cattle and buffalo and 100 million sheep and goats per year worldwide. In the United States $20 billion are lost annually from animal diseases. A 50 percent reduction in losses from epidemic and infectious diseases is a worthy and possible goal.

Substantial gains can come from developing genetically superior animals for low-quality diets, increasing the availability of lignin and cellulose (the major feedstuffs of many developing countries), improving the use of non-protein nitrogen for conversion by ruminants into protein, improving feeding and management at critical times in the life cycle, and eventually developing the rumen bypass for increased protein utilization and appetite stimulation.

Improved fertility will come from estrus synchronization or control of the reproductive cycle, semen preservation, improved pregnancy detection, multiple births, and superovulation. All can be used to increase the number of offspring per breeding unit. Low reproductive efficiency of livestock now results in 50 percent mortality. Twinning in dairy and beef cattle can become a reality through hormonal treatment for superovulation, controlled breeding, recovery of fertilized eggs, and nonsurgical embryo transfer to recipient females of lesser breeding value.

Controlled environments are a realm of technology scarcely tapped for livestock production. They are widely used now for poultry and swine, and are under test for lambs and beef cattle. Programmed lighting (photoperiod) and temperature regulation can have marked effects on crop productivity. Such experiments should be vigorously pursued to study their effects on productivity of dairy and beef cattle, feeding efficiency, rates of gain, hormonal relationships, and reproductive behavior.

Related Resources

Land, water, energy, and human labor are the basic resources needed for new crop and livestock technologies. They are interrelated and one can be partly substituted for another. Because of the recent decrease in the output/input ratio for energy into maize production in the United States, millions of hectares of land have been freed from maize production as a result of technological advances. Soybean hectarage expanded almost proportionately. Production of soybeans requires less energy from fossil fuels than maize does because of a built-in supply of nitrogen through biological nitrogen fixation, and less energy is needed for drying the harvested product.

Land, water, and energy interrelationships are readily apparent in irrigated areas, where an additional energy input is required for water management. Center-pivot irrigation systems in Nebraska require 10 times the fossil fuels needed for all other production requirements of maize (Splinter, 1976). Fossil energy power can be used to reduce the labor-manpower inputs and vice versa. Also, increasing the intensity of land management (through the use of water, fertilizer, pesticides) can reduce the amount of land needed, and the reverse is true. The land and water base can also be changed with time and technology.

Two major problems are now restricting the natural substitution of energy, labor, and land in agricultural production: (1) prime agricultural land is disappearing worldwide because of irreversible use, and (2) the land is being seriously degraded by erosion.

Problems of soil erosion are global. For 40 years the Soil Conservation Service of the U.S. Department of Agriculture has promoted sound conservation practices, supported by technical and financial assistance, education, persuasion, and appeals to good land stewardship. These efforts have not been very effective. After 40 years, no more than 25 percent of our farm lands are under approved conservation practices. Meanwhile, enormous losses of topsoil from our best lands continue to occur (8.3 tons per hectare per year). The situation is worse in tropical areas (Eckholm, 1976; Pimentel et al., 1976).

New energy-efficient land- and water-utilization strategies can be designed, and conservation technologies can be developed, to increase yields of small and large holdings. The introduction of resource conservation on farms of all sizes would exploit one of the greatest opportunities to improve the world's food production. In the United States, elaborate agricultural technologies have been developed using labor-saving equipment, on large-scale units, and with huge inputs of capital, management, and resources. Much of the world may not want or be able to handle such technologies in areas where energy, land, and water are in short supply, unemployment is rampant, and management is lacking. Food-producing technologies should be sought that are scale neutral, labor intensive, and result in stable production at high levels and with a minimum of capital, management, and resources. The U.S. scientific community has had little experience in this type of research. There are, however, some examples.

Reduced tillage is one. The plow, long a symbol of American agriculture, has been responsible for insolvable losses from wind and water erosion. Through the use of appropriate chemicals for weed control, improved crop productivity is possible without plowing the land. The plow is gradually being retired. In 1976 in the United States, 2.9 million hectares were planted without tillage, and on over 21 million additional hectares tillage was reduced from the conventional level. Reduced tillage is the most significant technology man has yet developed to control soil erosion, maximize cover on the land, and conserve energy, labor, water, soil fertility, and organic matter in the main food-producing areas of the world (Triplett and Van Doren, 1977). In addition, a greater share of land in hilly areas can be brought into production or planted in more profitable crops.

The number of kilograms of water needed to produce a kilogram of food is an important variable. Practically no research has been done in this area. Transpiration losses among species vary on a scale of 100 for pineapple, 400 to 500 for cereals and seed legumes, and over 1,000 for some fruits and vegetables. The water requirement of sugarcane per unit of land in Hawaii is five times that of pineapple. Why these differences? Research can provide the answers.

Equally important are new water-management technologies. Efficiency in using irrigation water varies from 20–40 percent in the United States to 80–85 percent in Israel. Drip or trickle irrigation, first developed for large-scale crop production in Israel, reduces by 50 percent the water now used in conventional irrigation systems (flooding, sprinkling, furrow) for food-crop production. There are many other advantages of drip irrigation for high-value crops. Soil erosion is reduced to a minimum; no land is wasted to build irrigation ditches; and there is no leaching, run-off, or drainage water pollution. The most effective weed-control methods and fertilizers can be used, and crop-management and harvesting operations can be conducted without interference. High-frequency irrigation shows even greater promise for efficient water use (Rawlins, 1976).

Farming Systems

Technologies relating to farming systems are the most promising of all for enlarging the supply, increasing the dependability, and improving the nutritional quality of food. They are also the most difficult to design and execute. Research on management of farming systems is contrary to the traditional disciplinary approach. Modern agricultural technologies are designed to improve a single crop or animal species rather than an existing stable system in total. Special and expanded administrative skills are required. Future progress in increasing crop and livestock productivity through research must come as a result of careful coordination of simultaneous efforts on several frontiers.

The number of subsystems at the farm level is enormous. Historically in the United States and currently in developing nations, the typical small farm is a complex system. It uses animals for power, fuel, and food, and raises many kinds of plants for food, feed, fuel, and for sale (American Society of Agronomy, 1977). There are diverse systems dealing with soil, water, energy, fertilizer, mechanization, labor, waste, pests, and climate. There are cropping systems, livestock systems, and tillage practices which span all of these resource systems. There are many alternative systems for managing beef cattle operations, swine units, dairy herds, and the production of corn, cassava, cotton, soybeans, and sugar beets. Farming technologies must be broken down into local geographical and ecological areas, and to site-specific socioeconomic, food-policy, institutional, and political conditions.

Food systems must extend beyond the farm gate. There are processing, handling, storage, transportation, packaging, and marketing systems. Domestic and international trade and port facilities are involved. There are regulatory matters, taxes, credit and subsidies, institutions to serve and create new technologies, a complex of public and private support, food policies, and an infrastructure to provide supplies and services. The ultimate goal for the producer is stable production at high levels with greater efficiency in resource use, and the development of management strategies that will increase the margin of profit.

CRUCIAL ISSUES

The success of any food-production and delivery system resides not only in the application of science and new biophysical technology but in economic, social, and political incentives to produce, based on a free-enterprise system and a family farm base. Prices are the key signals for farmers to produce. Unless farmers can expect a profit, their production will slow down and often cease beyond what is required for their families. The profit incentive is the right to plan, to earn, and to enjoy the fruits of labor.

Half the people of the world depend on subsistence agriculture, and 40 percent of the cultivated land of the world is in the hands of subsistence farmers. Government regulations and political constraints frequently establish prices and control profits. There is often little incentive to produce beyond what is needed for immediate survival.

Storage capability of indigenous production in subsistence agriculture is grossly lacking, but if available would provide production incentives for farmers, family food security, opportunity to manage a part of the market system, and a means of preserving seed stocks.

Technologies addressed to small-scale, labor-intensive, capital- and resource-scarce farming would capitalize on an unexploited food production frontier. The small farmer has not, as yet, generally benefited from modern agricultural technology. "Intermediate" technology is the prerequisite.

The income of a billion people on this earth depends on farms of 5 hectares or less: 75 percent of farms in tropical Asia are smaller than 2 hectares; 69 percent of farms in Central America are less than 5 hectares; and the average size of farms in 20 tropical African countries is only 5.4 hectares. The output per hectare on these miniholdings can exceed by one-third that of large U.S. farms. Increments of fertilizer or pesticides or other technologies, or economic and social incentives, in much of Southeast Asia, Latin America, and Africa, would have telling effects on food productivity.

South Korea during 1976 became 95 percent self-sufficient in the production of rice, its number one food crop. Technology made its contribution by introducing "Tongil," a new dwarf type of rice that is cold tolerant, early to mature, and resistant to "blast." Equally important, the farmer was encouraged by government policies and a strong market demand that resulted in a price increase of 121 percent.

Technologies that would improve yield stability of food crops could be as important as an increase in yields in encouraging farmers to adopt new practices. An example is pest control. Without a sense of stability, farmers are not likely to make an investment in agriculture that requires more than one growing season to amortize.

Special problems face the United States in its current assessments of food production technology. The time between a basic research discovery and its application averages 13 to 19 years. The time from introduction of a new technology until its adoption reaches the expected ceiling is 35 years. It takes 6 to 10 years to train scientists to do research. Crop and livestock protection technology becomes obsolete at an alarming rate. New crop varieties have an average life of only 5 years. Increasing expenditures (over 10 percent of the total) are going into defensive research, largely nonproductive for the creation and advancement of new and improved technologies.

Overriding all other considerations is the availability of scientists. An overwhelming majority of the trained agricultural scientists in the United States come from land-grant universities and state colleges with affiliated state agricultural experiment stations and cooperative extension services. These institutions, with one or more in each of the 50 states, are the agricultural and food research manpower reservoir of the nation.

TABLE 1. Enrollment in U.S. agricultural institutions.

Year	Enrollment
Baccalaureate Degree Programs	
1963	34,952
1970	54,348
1971	58,779
1972	64,845
1973	72,644
1974	81,736
1975	92,161
1976	98,183
Undergraduate Women	
1973	14,576
1976	27,700

This reservoir is being seriously eroded. Student enrollments have almost tripled since the mid-1960's (Table 1). This growth in numbers has been accompanied by a great diversity in educational background and experience. Meanwhile, there has been little or no increase in agricultural experiment station faculty or resources, or in the number of agricultural scientists in the Agricultural Research Service of the U.S. Department of Agriculture (Table 2). These two pools of manpower, with their supporting staffs, are the chief human resources in the creation of new food-production technologies.

Increases in teaching faculty to handle the greatly expanding student enrollments have, because of budgetary constraints, fallen far short of the escalating teaching loads. As a result, professional faculty have had to shift some of their research time to teaching and student advising. Most faculty

TABLE 2. Scientist years in support of agricultural research in the United States.

Year	Number	Year	Number
1966	10,457	1971	10,119
1967	10,042	1972	10,391
1968	9,817	1973	10,420
1969	10,039	1974	10,397
1970	10,265	1975	10,447

in land-grant universities have joint appointments in teaching and research.

The erosion of research has also included equipment and facilities, much of which is jointly shared in teaching and research. Coupled with this has been a shortage of new equipment and improved facilities during the last decade of unprecedented growth in enrollment. The availability of new technological equipment is far ahead of the ability of scientists to purchase it. No federal funds for improved facilities in the state agricultural experiment stations have been allocated in 10 years.

New international commitments relating to Title XII "Famine Prevention and Freedom from Hunger," Foreign Assistance Act of 1975, will put a further strain on those knowledgeable in the technologies relating to food production (Wharton, 1976).

The frontiers for creating and assessing new technology in food production must be pursued by scientists. Motivation and incentives are important for both scientific discovery and food production.

THE CHALLENGE

There is a degree of urgency in assessing technologies for food production. The number of alternatives available for research is staggering. They range from labor-saving technology for capital, resource, and management intensive operations to improved storage and farming systems for optimizing indigenous production in subsistence agriculture. Meanwhile, the world is growing by 70 million persons a year.

The world food problem is not one of production capability. It is one of delivery. The food has to be where the people are, and they must have money to buy it. Only poor people have a problem obtaining food.

The United States stands as the bastion against hunger, with a monopoly on the world's surplus food far exceeding that of any nation in recent history. Never has any country exported so much food or been so generous. Current agricultural exports exceed those of any other commodities, and approached $23 billion in 1976. These exports contributed heavily to our national economy and a favorable international trade and balance of payments. Of the food supplied in recent years to starving nations, 84 percent has come from the United States.

Research on the biological processes that control and now limit crop productivity has been emphasized in this chapter. This kind of research has the greatest potential for increasing world food output, and greater scientific inputs are critically important for the United States.

Farm yields rose rapidly in the United States in the 25 years following World War II (Figure 1). These were accompanied by massive inputs of

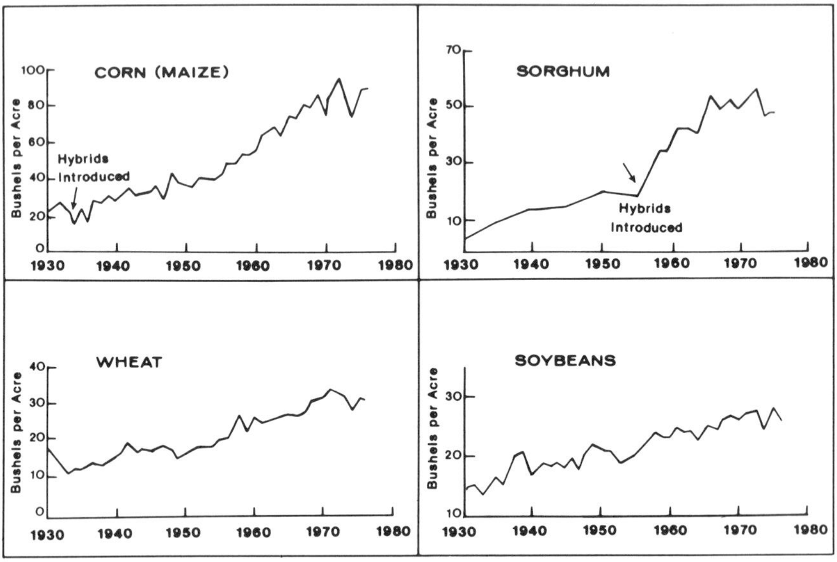

FIGURE 1
Yields in bushels per acre in the United States, 1930–1976.

fertilizer, pesticides, machinery, and capital. There was plenty of food at low cost. The result was that the average U.S. expenditure of income on food dropped to an all-time low of 16.3 percent in 1972, even though more protein and other nutrient-rich processed foods were consumed.

A dramatic change has occurred since 1970. Yields of all major food and feed crops (corn, soybeans, wheat, sorghum, potatoes) in the United States have reached a plateau and even fallen off in some cases (Figure 1). World grain yields have declined since 1970 (Figure 2). Prices for farm products in the United States have become more volatile and uncertain (Figures 3 and 4). The extreme variations in Soviet grain production (Figure 5) are partly responsible. Meanwhile, energy-intensive farm inputs have risen sharply and continue an upward trend. This has placed American agriculture and the farmer in a vulnerable position, with no historic precedent.

Increased profitability for the farmer, on the one hand, and reasonable food prices for the consumer, on the other, are both predicated on sustained increased production. These two complementary objectives underscore the urgency for greater investments in high-priority, mission-oriented basic research to increase stability and improve food production.

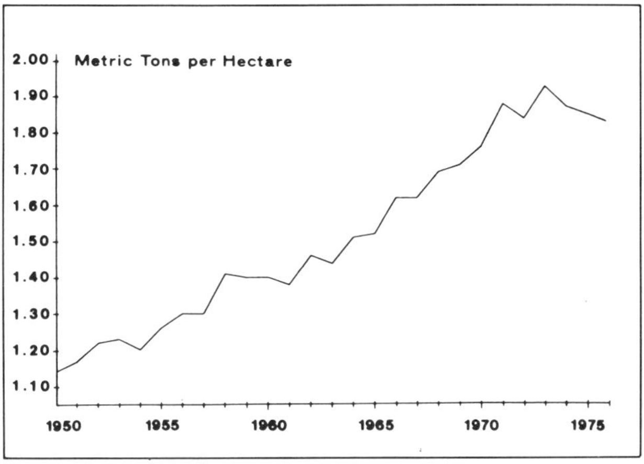

FIGURE 2
World grain yields, 1950–1975.

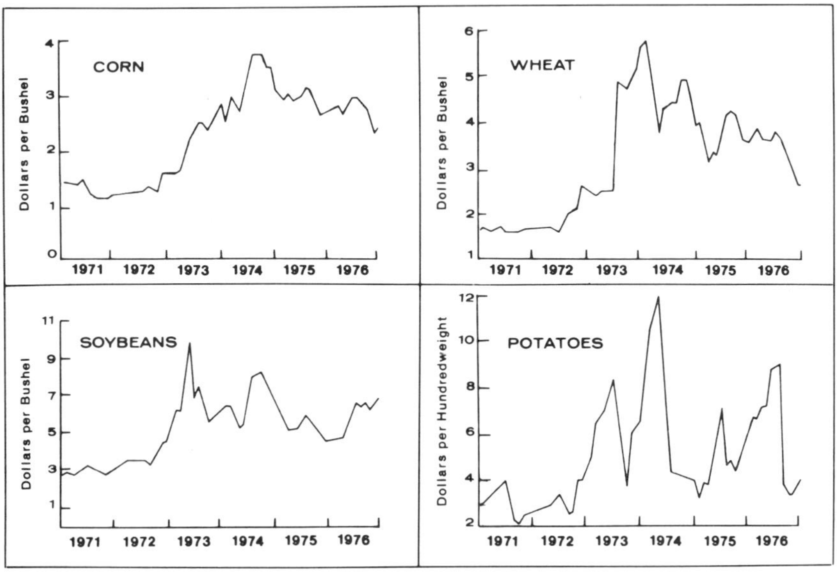

FIGURE 3
Price fluctuations of major food commodities (corn, wheat, soybeans, potatoes) in the United States (Chicago Board of Trade), 1971–1976. Potato prices are a composite of all Michigan prices.

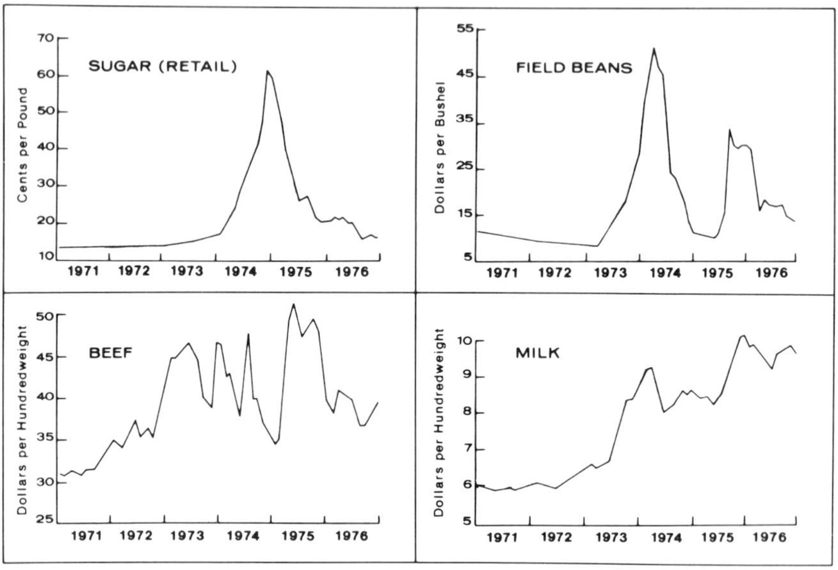

FIGURE 4
Price fluctuations of major food commodities (sugar, field beans, beef, milk) in the United States, 1971–1976.

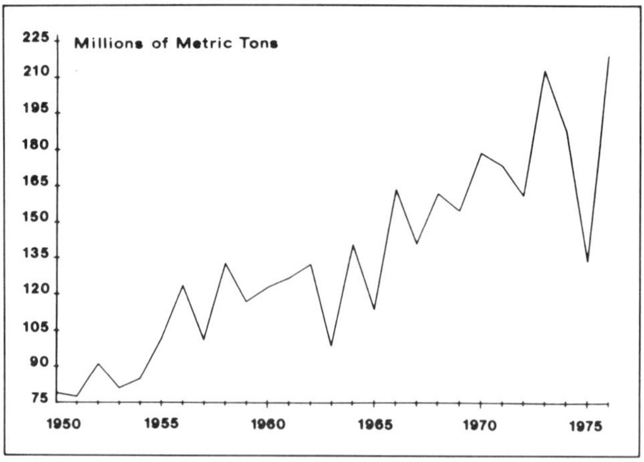

FIGURE 5
Grain production in the USSR, 1950–1976.

References

American Society of Agronomy. 1977. *Multiple cropping*. Madison, Wis.

Byerly, T. C. 1977. Ruminant livestock research and development. *Science* 195:450–456.

Committee on Agricultural Production Efficiency. 1975a. *Agricultural production efficiency*. Board on Agriculture and Renewable Resources, Commission on Natural Resources, National Research Council, National Academy of Sciences, Washington, D.C.

———. 1975b. *Enhancement of food production for the United States*. Board on Agriculture and Renewable Resources, Commission on Natural Resources, National Research Council, National Academy of Sciences, Washington, D.C.

Committee on Climate and Weather Fluctuations and Agricultural Production. 1976. *Climate and food*. Board on Agriculture and Renewable Resources, Commission on Natural Resources, National Research Council, National Academy of Sciences, Washington, D.C.

Crop productivity-research imperatives. 1976. Proceedings of the International Conference jointly sponsored by the Michigan Agricultural Experiment Station and the Charles F. Kettering Foundation.

Eckholm, E. P. 1976. *Losing ground: Environmental stress and world food prospects*. W. W. Norton and Co., New York.

National Academy of Sciences. 1975. *Contemporary pest control practices and prospects*. Report of the Executive Committee Study on Problems of Pest Control, Environmental Studies Board, National Research Council, Washington, D.C.

Pimentel, D., E. C. Terhune, R. Dyson-Hudson, S. Rochereau, R. Samis, E. A. Smith, D. Denman, D. Reifschneider, and M. Shepard. 1976. Land degradation: Effects on food and energy resources. *Science* 194:149–155.

Rawlins, S. L. 1976. High frequency irrigation and green revolution food production. Paper presented to the New York Academy of Sciences, December 1976, Philadelphia, Pa.

Science. May 1975 food issue. 188:503–651.

Scientific American. September 1976 food and agriculture issue. 235(3):31–205.

Shanmugam, K. T., and R. C. Valentine. 1976. Solar protein. *California Agriculture* 30(11):4–7.

Shapley, D. 1977. Will fertilizer harm ozone as much as SST's? *Science* 195:658.

Splinter, W. E. 1976. Center-pivot irrigation. *Scientific American* 234(6):90–99.

Triplett, G. B., Jr., and D. M. Van Doren, Jr. 1977. Agriculture without tillage. *Scientific American* 236(1):28–33.

U.S. Congress. 1976. *Assessment of alternatives for supporting high priority basic research to enhance food production*. Report of the Office of Technology Assessment Panel. Washington, D.C.

Wharton, C. R., Jr. 1976. Ecology and agricultural developments striking a human balance. Keynote address, Annual Meeting, American Society of Agronomy, Houston, Texas, November 1976.

Wittwer, Sylvan H. 1974. *Research recommendations for increasing food, feed, and fiber production in the U.S.A*. National Science Foundation, Washington, D.C.

———. 1975. Food production: Technology and the resource base. *Science* 188:579–584.

———. 1977a. The challenge to science for food production. Paper prepared for the U.S. Information Agency, Washington, D.C., January 1977.

———. 1977b. Increased crop yields and livestock productivity. Pages 66–135 in Marylin Chou, David P. Harmon, Jr., Herman Kahn, and Sylvan H. Wittwer, *World food prospects and agricultural potential*. Praeger Publishers, New York. Published in cooperation with the Hudson Institute, New York.

4 Assessment of technology in fiber production

George R. Staebler

Current technology might be rated on a scale of 1 to 10—if we were to take a literal approach to the subject of this chapter—and such a measure might range from "primitive" at one end to "advanced or sophisticated" at the other. Definitions, qualifications, and interpretations would be called for, and we would have to struggle with opinions in the absence of facts and figures. Probably we would bog down in obscure semantics, and the scale rating would not be one whit different from conclusions reached by gut feelings.

My gut feeling is that the current level of technology used in growing trees, compared with agricultural technology, would rate fairly close to the primitive end of the scale. Current technology is not totally primitive, however, because some forests are consciously regenerated: if naturally, by some means other than accidental ones; if artificially, by planting nursery-grown stock—a process requiring technology.

On the other hand, if the farmer plants hybrid seed in a carefully prepared seedbed and applies fertilizer at a rate determined by the needs of the hybrid crop he is raising and by the fertility of the soil he is farming, if he controls weeds with specially formulated herbicides, if he irrigates, if he harvests when the sugar or moisture content of his crop indicates maturity, if he uses machines designed and manufactured specifically for his crop, then one would no doubt rate the technology he uses as advanced and perhaps even sophisticated. By any standard, forestry has a long way to go to achieve agriculture's level of technology. But from another point of view, someone might have concluded a couple of centuries ago that a farmer who owned Thomas Jefferson's new invention, the moldboard plow, was using advanced technology simply because he was using the latest technology available. I submit that forestry does not come off well when compared with agriculture because its technology—although thoroughly proven—is too little used in raising tree crops.

Viewed still another way, one might say that the level of technology used is equal to the actual yield of a crop divided by the yield possible with full application of known technology. If proven cultural practices applied to a given soil type in a given area of the country could result in an average production of X for some crop, but actual production is one-half of X, we have to conclude that the effective level of technology is low.

The problem has been studied recently by the Panel on Biological Productivity of Renewable Resources Used as Industrial Materials, one of the six CORRIM panels (Committee on Renewable Resources for Industrial Materials, 1976). The panel drew its wood fiber data almost exclusively from *The Outlook for Timber in the United States,* the 1970 appraisal of the timber situation made by the Forest Service (U.S. Department of Agriculture, 1974).

The Forest Service estimated, from the findings of its forest survey, that the commercial forests of the United States were currently growing 2.7 cubic meters per hectare annually. They also estimated that if all forests were growing like normal forests of the type currently occupying the given area, and if the age classes were fully balanced (assuming harvest at the age of culmination of mean annual cubic increment), they would be growing at 5.2 cubic meters per hectare.

The actual increment divided by the potential increment here equals 52 percent. Although this figure is a possible expression of the current level of technology, it must be extensively qualified. The current annual increment at 2.7 cubic meters per hectare includes a lot of zero-growth hectares: those occupied by old growth with no net growth; very young stands, resulting from accelerated harvests in the past decade, which have not yet reached the grand period of growth; and the nonstocked hectares. Very likely these hectares are not balanced by the hectares of stands in their grand period of growth where current increment is greater than mean annual increment. Therefore, it is likely that current annual increment will continue to increase—even if technology remains unchanged—and that 52 percent is really an artifact applicable only to the 1970's.

The Outlook for Timber, in fact, bears this out in its estimate of production for the year 2000 at 2.8 cubic meters per hectare (up from 2.7) after accounting for projected removals. With a continuation of the 1960's level of management, our "level of technology" would go from 52 percent to 54 percent as a result, presumably, of achieving a better distribution of age classes, and as a result of the compounding effect of applying the current level of management, but with no significant improvement in technology.

Up to this point we have looked at the numerator of the fraction—the current annual increment over potential increment—as a measure of the status of technology. We should now look at the denominator, the potential increment. This is the figure that the Biological Productivity panel attempted to assess. The figure of 5.2 cubic meters per hectare assumes that each hectare's potential increment is equal to the maximum mean annual increment that appears in normal yield tables for the forest type occupying the hectare (or, on nonstocked hectares, the forest type best suited to the land).

Normal yield is the net yield of a fully stocked, even-age stand, regenerated naturally, with no cultural treatments applied between regeneration and harvest. Aside from the technology required to achieve full stocking through natural regeneration, and to prevent catastropic losses (as opposed to endemic losses), this is tantamount to no technology at all. We have to conclude that our denominator for full production is smaller than it should be, and 54 percent as our measure of technology applied is generous.

The Biological Productivity panel attempted to calculate a more realistic denominator—realistic in the sense that it can be attained with the application of current technology.

Even though I have just stated that achievement of a normal forest requires a minimum of technology, it is unlikely that normal yields will be achieved by consciously establishing normal forests. In any event, achievement of normal yields by this route would have to await replacement of all nonnormal forests. The question becomes: What yields would be possible by the year 2000 with aggressive application of known technology?

PROSPECTS FOR INCREASING YIELDS

The panel considered the opportunities for increased yields as presented by Ostrom and Gibbs (1973), who identified eight technological approaches: (1) Improve the site through cultivation, fertilization, drainage, and irrigation. (2) Convert existing forest types to faster-growing species. (3) Improve stocking through reforestation. (4) Introduce faster-growing genotypes within a given species. (5) Stimulate faster growth through weeding. (6) Recover a larger proportion of gross increment through thinning. (7)

Reduce losses to fire, insects, and diseases through better forest protection. (8) Increase allowable cut as a result of management decisions.

The panel's method was to assess the effects of the more significant of the eight approaches (or perhaps those for which estimates could be made and defended) in terms of growth per hectare, then multiply this average by the area of commercial forest land projected for the year 2000.

Improving the Site

The panel said that the productivity of 1.90 million hectares in the Douglas-fir Region could be raised by the equivalent of one site class through nitrogen fertilization. For southern forest types, the site quality of 2.35 million hectares of moist slash and loblolly pine types could be similarly raised by the application of phosphate fertilizer and by drainage. A new area-by-site class distribution resulted, the net effect of which was an increase in the productivity of the Douglas-fir Region of 5.5 percent, and of the Southern Pine Region of 3.2 percent. Considering that the area involved in this calculation is only 2.2 percent of the commercial forest land base of 192 million hectares projected for the year 2000, the panel can hardly be accused of stargazing.

Conversion of Forest Type

It should be apparent to careful observers of the American forest scene, even if they never get off the paved highways, that an appalling amount of America's cut-once forest is now a very inferior forest. This is true for hardwoods of desirable species and hardwoods of less desirable, or at least less productive, species as well as for softwood forest types and greatly inferior hardwood types of minimal productivity. The panel attempted to evaluate the southern forest opportunity only, and projected that one-half of the mixed oak-pine forest, on sites III, IV, and V, would be converted to the faster-growing pine and that one-tenth of the upland hardwood types on the same sites would be so converted. It was assumed that the 8.1 million hectares thus treated would be raised in productivity by one site class, resulting in a 40 percent increase in the growth projected for the South by *The Outlook for Timber.*

Reforestation

It is in the area of reforestation that the panel concluded the opportunity mainly exists to establish normal stands, or stands that produce like normal stands. After upgrading the sites by fertilization and drainage and by converting them to higher-producing types, and evaluating the effects of both practices by adjusting the distribution of area by site class, it was assumed that 90 percent of normal yield could be obtained on sites I and II, 80 percent on site III, 70 percent on site IV, and 60 percent on site V. The reductions are in line with the observed drop in yields over broad areas as opposed to those measured on small, selected sample plots, and also reflect the conviction that the highest order of management will be practiced on higher class sites and a lower order on the poorer sites. It was assumed that all of the 1.6 million hectares of currently nonstocked sites I, II, and III land would be reforested plus 1.2 million hectares (of 2.4) of site IV land and 0.4 million hectares (of 4.5) of site V. This constitutes a real increase in producing forest area.

Faster-growing Genotypes

Tree-improvement programs are well advanced in the West and South. The panel recognized this developing technology by increasing the projected normal yields (after adjustment for management intensity) by 15 percent on sites I and II, 10 percent on site III, and 5 percent on site IV (no gains are predicted for site V) in the Douglas-fir and Southern Pine Regions only. Although considerably less than the theoretical maximum, the added yield contributes 3.3 percent to projected yields for 192 million hectares of commercial forest land.

Weeding

No evaluation of gains from weeding was made, except as it contributes to normal yields.

Thinning

Few established technologies are as little employed in forestry as thinning —thinning as a cultural practice to increase the yield of even-age stands. The theoretical basis has long been established and verified through research, but for a number of reasons thinning has not been extensively adopted in wood-rich countries of the world. There can be little doubt that some or all of normal mortality in normal stands can be captured through an aggressive thinning regime. The Biological Productivity panel projects that the gains would be similar to those resulting from genetic programs: 15 percent on sites I and II, 10 percent on site III, and 5 percent on site IV, again in just the Douglas-fir and Southern Pine Regions.

TABLE 1. Projected changes in area of commercial forest land under intensive management, by site class and species group.

Forest Land Use	Site Productivity Class									
	I and II		III		IV		V		Total	
	million acres (million hectares)									
Softwoods										
1970	34	(13.7)	60	(24.3)	94	(38.0)	54	(21.9)	242	(97.9)
Projected, 2020	38	(15.4)	75	(30.4)	101	(40.9)	47	(19.0)	261	(105.6)
Hardwoods										
1970	16	(6.5)	54	(21.9)	95	(38.4)	67	(27.1)	232	(93.9)
Projected, 2020	15	(6.1)	50	(20.2)	78	(31.6)	58	(23.5)	201	(81.3)
Nonstocked										
1970	2	(0.8)	2	(0.8)	6	(2.4)	11	(4.5)	21	(8.5)
Projected, 2020	0	(0.0)	0	(0.0)	3	(1.2)	10	(4.0)	13	(5.3)
Total										
1970	52	(21.0)	116	(46.9)	195	(78.9)	132	(53.4)	495	(200.3)
Projected, 2020	53	(21.4)	125	(50.6)	182	(73.7)	115	(46.5)	475	(192.2)

Forest Protection

In 1970, according to estimates in *The Outlook for Timber,* 126 million cubic meters of wood were destroyed by fire, insects, disease, and storms. The panel, by starting with normal yields as a base, in essence assumes no losses, since normal yields are measured in stands which have not suffered catastrophic losses. Recognition that there will be losses, however, is inherent in the reductions in normal yields previously described as management-intensity adjustments.

Allowable Cut Effect

The concept that the current harvest of an unregulated forest can be tied to the growth initiated each year through cultural practices was discussed by the panel, but no adjustments were made for it. The concept is that the mean annual increment at harvest of a stand can be considered as part of the allowable cut on the day the stand is established. Increases in current growth brought about by cultural practices in existing stands (fertilization is the least complicated example) may also be included. This is in contrast to the allowable cut under sustained yield, which matches the harvest to the current annual increment of the existing unregulated forest, or under controlled straight area regulation. The main effect would be to hasten the day when full or increased production would arrive and to stimulate it by the economic incentives inherent in the concept.

The eight technologies discussed above would greatly increase the productivity of America's commercial forest land if they were widely practiced.

FUTURE FOREST LANDS

The Outlook for Timber projects that the commercial forest land of the United States will cover 196 million hectares in the year 2000 and 192 million in 2020 (see Table 1). The panel believes that the total will be 192 million hectares in 2000 and has based its projections on that figure. A summary of this is best stated by quoting the panel report:

> To obtain the projected volume of timber produced, the number of acres in each category is multiplied by the assumed production per acre per year. If the total area was fully-stocked so that yield table predictions could actually be produced, it was assumed that the production per acre would be the mid-point in each site productivity class. . . . As previously discussed, however, yield table values cannot actually be achieved over large acreages. We assumed, therefore, that management would be

TABLE 2. Projected productivity of U.S. commercial forests under intensive management, by site class and species group.

	Site Productivity Class									
	I and II		III		IV		V		Total	
Projected Growth, 2000	cubic feet per acre per year (cubic meters per hectare per year)									
Normal Yield	161.5	(11.3)	102.5	(7.2)	67.6	(4.7)	35.0	(2.4)		
Management Intensity	90%		80%		70%		60%			
Projected Yield	145.0	(10.1)	82.0	(5.7)	47.0	(3.3)	21.0	(1.5)		
Projected Production, 2000	million cubic feet per year (million cubic meters per year)									
Softwoods										
U.S.	5,510	(154.3)	6,150	(172.2)	4,747	(132.9)	987	(27.6)	17,394	(487.0)
Added Douglas-fir	602	(16.9)	128	(3.6)	30	(0.8)	0	(0)	760	(21.3)
Added Southern Pine	344	(9.6)	360	(10.1)	215	(6.0)	0	(0)	919	(26.0)
Total	6,456	(180.8)	6,638	(185.9)	4,992	(139.7)	987	(27.6)	19,073	(534.3)
Hardwoods	2,175	(60.9)	4,100	(114.8)	3,666	(102.6)	1,218	(34.1)	11,159	(312.5)
Total										
U.S.	7,685	(215.2)	10,250	(287.0)	8,413	(235.5)	2,205	(61.7)	28,553	(799.5)
With Added DF & SP	8,631	(241.7)	10,738	(300.7)	8,658	(242.3)	2,205	(61.7)	30,232	(846.8)

> sufficiently intensive to produce [the fractions of normal yield discussed above].
>
> Applying these assumptions, we obtain a mean annual production of 145 cubic feet per acre [10.1 cubic meters per hectare] per year on Sites I and II to 21 [1.5] on Site V [see Table 2]. Multiplying these values by the appropriate acreage projections, we estimate a potential productivity of softwoods of over 17 billion cubic feet [0.5 billion cubic meters] per year and hardwoods of over 11 billion [0.3 billion] for a total of 28.5 billion cubic feet [0.8 billion cubic meters]. This estimated productive potential is twice that of 1970 consumption of 14 billion cubic feet [0.4 billion cubic meters]. . . .
>
> Applying the potential gain in growth from thinning and genetic improvement, we predict an increase in productive potential of 760 million cubic feet [21.3 million cubic meters] annually in the Douglas-fir region and 1,119 million cubic feet [31.3 million cubic meters] from southern pine. Adding these two gains to the earlier estimate, we predict a total biological productive potential of . . . over 30 billion cubic feet [0.8 billion cubic meters] for all species.

This amounts to a potential growth of 4.5 cubic meters per hectare, considerably less than the 5.2 given as the normal yield potential. Our index of current level of technology becomes 2.7 divided by 4.5, or 60 percent, and appears more inflated than it did with the denominator that we said implied minimal technology.

I believe the panel report correctly assesses that the projected yields are realistic, and that they can and probably will be achieved. At the same time, there is ample room for improvement, by applying developed technology more widely in the commercial forest land, and by continued progress toward theoretical maximums.

By and large, however, we have come up with little to refute our initial gut feeling that the current level of technology practiced in American forestry is only a little way from the primitive end of the scale. The opportunity and the challenges are great.

References

Committee on Renewable Resources for Industrial Materials. 1976. *Biological productivity of renewable resources used as industrial materials*. Board on Agriculture and Renewable Resources, Commission on Natural Resources, National Research Council, National Academy of Sciences, Washington, D.C.

Ostrom, C. E., and C. B. Gibbs. 1973. Biological potential for and biological constraints on timber yield. U.S. Department of Agriculture, Forest Service (draft manuscript).

U.S. Department of Agriculture, Forest Service. 1974. *The outlook for timber in the United States*. Forest Resource Report 20.

5 Energy requirements in managing for food and fiber production

William E. Splinter

The key to the American food machine, wherein each farmer produces sufficient food for himself and 50 others, has been the utilization of fossil fuels through mechanical systems which greatly magnify individual productivity. Agricultural tractors, combines, crop driers, irrigation systems, fertilizers, and pesticides are all manifestations of the use of fossil fuels, primarily natural gas and petroleum. Fiber production in the form of cotton is similar to crop production in mechanization and extensive use of chemicals. Fiber production in the form of pulpwood has seen major shifts to mechanization in the past 10 years, with the introduction of chain saws, tree shears, and mechanical transplanters and loaders.

Petroleum, which supplies 45 percent of the total U.S. energy and 25 percent of the energy for food production (Brier, 1977), peaked in production in the United States in 1970, and production has declined about 2 percent per year since then. Estimates are that we have from 23 to 28 years of reserves of oil left in the United States (excluding Alaska) at our present rate of production. Natural gas reserves, which supply 33 percent of U.S. energy and 40 percent of energy for food production, may be essentially depleted in the United States within 30 years at present levels of production, since proven reserves have been decreasing about 4 percent per year since 1970.

The cost of oil from shale or the tar sands is estimated to be $20 per barrel based on 1976 prices ("World Oil Crunch Seen in the 1980's," 1977). Synthetic crude from coal would cost $22 per barrel, and synthetic natural gas from coal would cost $4.50 per quad Btu's, compared with $1.42 for new gas at the wellhead today.

World population, on the other hand, is growing at a rate of 1.9 percent per year and will double in 36 years to about 8 billion people. Since we export about the same amount of foodstuffs as the U.S. population consumes directly, rising energy costs for American food production will have worldwide impact.

ENERGY CONSUMPTION IN THE FOOD CHAIN

Steinhart and Steinhart (1974) claim that the fossil fuel energy used in the food chain has grown to exceed the energy return in the food consumed in the United States. Booz et al. (1976) estimate that 16.5 percent of the total U.S. energy budget is invested in food production, distribution, and preparation. Direct energy consumption on the farm accounts for only 1 percent of the total U.S. energy budget, but energy inputs in the form of chemicals and transportation increase this to 2.9 percent. Food processing adds 4.8 percent, the distribution system 1.7 percent, commercial food service 2.8 percent, and home food preparation 4.3 percent. Fiber crops and forest products account for an additional 5.5 percent.

An estimate of the energy used in the various on-farm operations is shown in Table 1 (Economic Research Service, 1974) for the United States. The indirect use of energy through fertilizers is the largest single energy investment. One expects field operations to be a major energy investment; but transportation, which includes on-farm as well as farm-to-market expenditures, uses only slightly less energy. Irrigated agriculture, which accounts for only 2 percent of the total cropland, uses a significant amount of energy in pumping water, either directly from wells or by lifting surface water to higher elevations. Energy for production of livestock is primarily in the form of feed mixing and processing. Crop drying on the farm is mainly accomplished through use of liquefied petroleum gas and consumes about the same energy as that used indirectly through pesticides.

The energy-use pattern for individual states varies greatly depending on cultural systems. In Nebraska, which has some 2.5 million hectares under irrigation (2.1 million hectares of which are irrigated by water pumped from deep wells), irrigation is the second largest energy investment, as shown in Table 2 (Bashford et al., 1977). Irrigation is used primarily for grain crops such as corn and sorghum, consequently energy use for on-farm crop drying is relatively high in Nebraska.

TABLE 1. Energy use in production agriculture in the United States (1974).

Energy Use	Joules x 10^{12}	Percent
Fertilizer	648	30
Field Operations	416	20
Transportation	342	16
Irrigation	284	13
Livestock	168	8
Crop Drying	121	5
Pesticides	110	5
Miscellaneous	63	3
Total	2,152	100

Considerable attention has been given to corn production—with some justification, since it is our most important grain crop. Three independent studies, which have estimated energy budgets for corn production, including machinery and chemical inputs, are compared in Table 3. Although the totals are close, there are disparities between individual input estimates. It would therefore be constructive to discuss these differences.

Neither Pimentel et al. (1973) nor Heichel (1976) included the stover weight as a component of yield. In general, the weight of the kernels is about 43 percent of the total above-ground plant weight. This stover is used by many farmers as livestock feed through pasturing, at about a 50 percent recovery level, and has further value as a source of nutrients and as a soil amendment to improve tilth.

TABLE 2. Energy use in production agriculture in Nebraska (1974).

Energy Use	Joules x 10^{12}	Percent
Fertilizer	46	33
Field Operations	23	17
Transportation	13	10
Irrigation	33	25
Livestock	6	4
Crop Drying	11	8
Pesticides	2	2
Miscellaneous	1	1
Total	135	100

If total dry-matter weight is included, the output/input ratio increases to 7 and 8 for the Pimentel et al. and Heichel estimates, respectively. The Roller et al. (1975) estimate of 12:1 probably establishes an upper limit for present technology. At any rate, all three studies show that corn production is, overall, an energy amplifier.

The three studies indicate conventional plowing-discing systems for soil preparation. Preliminary results from studies under way at the University of Nebraska show an energy investment of 2,380 million joules per hectare for a conventional system. Moldboard plowing consumes 13.75 liters of diesel fuel per hectometer, while chisel plowing cuts this in half. Going to minimum tillage systems, such as the Nebraska Till Plant System (Lane et al., 1973), eliminates plowing and at least one discing; and total fuel energy for tillage, planting, and harvesting is reduced to 1,020 million joules per hectare. Lane et al. estimated that only one-fifth of the corn farmers in Nebraska plowed their land in 1973.

Going to reduced tillage systems appears to require 1 to 2 kilograms per hectare of additional herbicide. Green and McCulloch (1976) report an energy cost of 460 million joules per kilogram to manufacture Paraquat and 190 million joules per kilogram to manufacture Atrazine. Heichel (1976) estimates that minimum tillage reduces total energy costs by 6 percent. Thus, reduced tillage systems offer the advantage of reduced energy requirements in addition to conservation of soil, for which they were originally developed.

There is considerable variation in the values reported for energy investments in agricultural tractors and machinery. This is probably one of the most difficult energy uses to estimate because of the complex nature of the manufacturing process and the variety of machines, ranging from harrows to self-propelled combines. Based on 1974 figures from a major equipment manufacturer, Doering et al. (1976) reported that hay and forage equipment required an energy investment of 26,200 million joules per metric ton to manufacture small grain planters, 35,900 million per ton for sprayers and secondary tillage equipment, 46,500 million per ton for primary tillage and large grain planters, 67,700 million per ton for harvesters, and 86,600 million per ton for tractors. Doering thinks that the Pimentel et al. estimate is about 25 percent high. Thus, a figure of 3,260 million joules per hectare might be considered a reasonably close estimate of the energy investment in manufacturing equipment for corn production. Minimum tillage systems eliminate the plow, but most other implements are still used. This is a minor reduction of about 6 percent in equipment weight and of about 5 percent in energy investment.

Fertilizers account for 30 percent of the on-farm energy use nationwide (Table 1). Hoeft and Siemens (1975) report that corn yields begin leveling

TABLE 3. Comparison of energy budgets for corn production (in millions of joules per hectare).

Input	Pimentel et al.* (1973)	Heichel (1976)	Roller et al.† (1975)
Fuel	8,245	2,017	3,075
Equipment Manufacture	4,345	7,563	1,765
Fertilizer	10,924	13,604	19,419
Pesticides	228	455	455
Seed	652	—	124
Labor	51	—	8
Irrigation	352	—	—
Drying	1,241	6,621 (Drying and Transport)	—
Transport	724		(incl. above)
Total Input	26,762	30,260	24,846
Output	84,500	104,300	318,600
Energy Output per Input	3.0	3.5	12.8

* Pimentel et al. included a figure for electricity which was prorated from total U.S. farm consumption. This would include dairying, feed processing, irrigation, and so forth. It is deleted here.

† The Roller et al. output total is the maximum potential total plant yield, dry-matter weight, corresponding roughly to a grain yield of 9.6 metric tons per hectare. The other two studies are based on grain yields of 5 and 6.3 metric tons per hectare respectively.

off at 150 kilograms per hectare of nitrogen and are essentially level beyond 200 kilograms per hectare in Illinois. Anhydrous ammonia, at 150 kilograms per hectare (roughly the values used by Pimentel et al. and Heichel), was calculated to have an energy cost of 9,656 million joules per hectare to manufacture, 428 million to transport, and 261 million to apply, for a total of 10,345 million joules per hectare. Urea solution costs 13,363 million joules per hectare to apply, and ammonium nitrate solution 12,938 million joules per hectare. Roller et al. (1975) used an application of 235 kilograms per hectare with correspondingly higher energy cost.

Although irrigation accounts for only 13 percent of the total energy used in agricultural production, it is critical to the agriculture of the 17 western states, and increasingly to many of the eastern states. For example, Florida has increased irrigated acreage by 101 percent, Missouri by 123 percent, and Minnesota by 682 percent over the past 10 years. During this period, Nebraska has experienced the greatest increase in irrigated land, 979,000 hectares, followed by Kansas with 584,000 hectares.

California, with the largest irrigated acreage, has over 91,000 wells plus several major irrigation projects which provide water from river diversions. Irrigation water, together with municipal water, is carried through major aqueducts having both electrical generating plants and electrical pumping plants. For example, the Central Valley Project handles some 10,645 million cubic meters of water generating 5,091 megawatts at a cost of 795 megawatts (Roberts and Hagan, 1975) for a net energy gain. The Bureau of Reclamation, Department of Interior, has installed a number of major surface irrigation projects in the 17 western states irrigating some 3,755,565 million hectares.

Most of the irrigation water from the Rocky Mountains east is supplied by pumping from deep wells, streams, and ponds. Texas has over 100,000 irrigation wells and Nebraska has over 61,000. Batty et al. (1975) have estimated the total energy cost for pump irrigation systems, including manufacturing energy, labor, earth work, operating energy, and expected life of components. Excluding pumping lift, surface irrigation without reuse of runoff water required an annual energy investment of 1.43 million joules per hectare to apply 915 millimeters of water. The addition of a reuse system, which recovers about 30 percent of the water applied, required an annual investment of 2.35 million joules per hectare. To conserve underground water, reuse systems are now mandatory in some areas of Nebraska and are highly recommended elsewhere. Center-pivot systems require 12.9 million joules per hectare. To these energy requirements must be added energy to lift water. For a 30-meter lift, applying 915 millimeters of water, 4,000 million joules per hectare would be needed. Wells in Arizona average 300 meters. Therefore, the total energy investment in irrigation application systems is generally much less than the energy required to lift the water.

Because of their adaptability to rolling terrain and sandy soils, center-pivot systems have been installed in large numbers throughout the United States and in various parts of the world. Satellite imagery-identification techniques indicate that over 11,750 of these systems were in operation in Nebraska during 1976. Table 4 shows an energy budget for a typical center-pivot installation in Nebraska where 3.4 cubic meters per minute of water are pumped from a depth of 30 meters against a pressure head of 545 kilonewtons per square meter, irrigating 54 hectares of land (Lane et al., 1973; Splinter, 1976). Over a season, applying 381 millimeters of water to the crop, this system consumes 496 liters of diesel fuel per hectare or its equivalent in electrical energy or natural gas. This is 10 times the energy used in the field operations, and the indirect energy use in the form of fertilizer is 5 times the energy used in the field operations.

Where irrigation is by surface systems rather than by high-pressure sprinklers, the energy cost is lower. For a typical surface system in Nebraska, again pumping from a depth of 30 meters, the pressure head is only

TABLE 4. Fuel energy budget of center-pivot irrigation for corn.

Activity	Billion Joules per Hectare
Plow, Plant, Cultivate, Combine	1.9
Fertilize (150 kilograms of nitrogen per hectare)	9.6
Apply Herbicide	1.5
Irrigate (pump from 30 meters)	18.8

60 kilonewtons per square meter. Even though most installed surface systems are not as efficient in applying water, requiring 508 millimeters for a season rather than 301, the total energy cost is still lower (about 280 liters of diesel fuel per hectare).

There is great economic pressure at this time to reduce pumping costs. In three surveys, pumping plants have been found to be poorly adjusted (Axthelm, 1962; Fischbach et al., 1967; Noffke et al., 1975). This results in operation at an average of only 65 to 70 percent of the Nebraska Performance Standard (Schleusener and Sulek, 1959). Fischbach (Lane et al., 1973) estimated that adjusting pumping plants to the Nebraska Performance Standard could save Nebraska farmers $3 million per year at 1973 prices for fuel, a figure which would be doubled at today's fuel costs.

Lowering the pressure head for sprinkler systems is receiving considerable attention at this time. Center pivots are now on the market which use low-pressure nozzles allowing a reduction in pressure head to as low as 120 kilonewtons per square meter. This would reduce pumping energy requirements by half. These systems, though economically attractive, apply water at an intense rate, creating potential problems in runoff and erosion on many soils which have low intake rates, steep slopes, or which tend to puddle, thereby negating the advantages gained through energy reduction.

For surface irrigation systems, typically 30 percent of the water applied runs off. Tail water can be recovered at low energy cost, and using it in automated gated pipe systems allows application of water at efficiencies at least as good as for center-pivot systems (Fischbach and Somerhalder, 1971). This improved application efficiency allows a proportional decrease in energy for pumping, and commercial development of this principle is now under way in several states.

A final method of reducing energy for irrigation comes about through scheduling of applications to match plant needs more exactly. The typical farmer, through experience, has found the economic consequences of overirrigation much less severe than those for underirrigation. Lacking precise knowledge of water availability in the soil and anticipated plant water demands, he must apply sufficient water to supply the crop with adequate water during peak water-use periods, even though these periods may be short in duration. Use of the computer to calculate evapotranspiration and available soil water, together with the use of resistance blocks to monitor soil moisture, has allowed close estimation of irrigation requirements. Gilley and Watts (1977) have stated that through pumping-plant adjustment, irrigation scheduling, and use of lower-pressure systems, Nebraska farmers can obtain comparable yields with two-thirds to one-half of the water and energy they are now using.

Trickle irrigation, which has received major use in the desert areas, requires an investment of 10.2 million joules per hectare. These systems are best adapted to tree crops and high-value vegetable crops because of the high initial investment.

Crop drying, which uses 5 percent of the total energy used in agriculture, can utilize the naturally lower humidities in the western United States to effect considerable energy savings. Lane et al. (1973) estimate that 20 percent of the energy used to dry corn in Nebraska could be saved by utilizing natural air drying for grains below 24 percent moisture.

Although grain crops give a positive net energy return, fruit crops are only marginal, and vegetable crops are generally deficit in energy return (Heichel, 1976). In the conversion of grain to meat protein through livestock, the agricultural production system has further losses.

An average steer in a feedlot on a finishing ration of high grain content produces 1 kilogram of live weight for 6.6 kilograms of dry-matter weight of grain. If we figure a 65 percent dressout of meat at 70 percent moisture content, we find that it would take 34 kilograms of dry-matter weight of grain ration to yield 1 kilogram of dry-matter weight of meat. For swine (which require 2.7 kilograms of dry-matter weight of grain per kilogram of live weight), at 60 percent dressout and 70 percent moisture meat, it would take 15 kilograms of dry-matter weight of grain per pound of dry-matter meat. For broilers (feed efficiencies are now on the order of 1.7 kilograms of dry-matter weight of grain per kilogram of live-weight gain), at 70 percent dressout and 75 percent moisture meat, it requires 9.7 kilograms of dry-matter weight of grain per kilogram of dry-matter meat. And for catfish, which are now produced commercially on thousands of farms, it requires 1.2 kilograms of dry-matter weight of grain per kilogram of live-weight gain, which at 65 percent dressout and 75 percent moisture gives a yield of 1 kilogram of dry-matter meat for 7.4 kilograms of dry-matter weight of grain. It would appear, therefore, that over the long run, as population pressures increase and as fuel supplies become more costly, the relative price of meat as a food source will increase and so will the percentage of grain in our diet. But meat will continue to be used as a food source in the foreseeable future, because livestock do convert roughage or low-quality feeds into food which can be used by man.

ALTERNATIVE ENERGY SOURCES

Roughly one-fourth of the feed for animals is voided as waste. In 1975 there were 43.7 million beef cows, 11.1 million dairy cows, 10.2 million feeders, 49.6 million swine, 13.3 million sheep, 280 million layers, 386 million broilers, and 49.7 million turkeys (USDA Statistical Reporting Service, 1976). Total residue voided would amount to 102 million metric tons (Van Dyne and Gilbertson, 1978). Waste from beef cows and sheep cannot be easily recovered, since these animals are primarily on pasture. However, the other animals are, for the most part, raised or fed where waste could be recovered. This would result in a potential for 52 million metric tons of recoverable manure.

If used for fertilizer, this waste would produce 3.8 million metric tons of nitrogen, 0.9 million metric tons of phosphorus, and 2.2 million metric tons of potassium. At 224 kilograms per hectare for nitrogen, this waste would fertilize 16.6 million hectares of corn, or about 45 percent of the total U.S. area in corn. Phosphorus and potassium are not readily lost, but nitrogen dissipates easily as ammonia. Average nitrogen losses in handling and storage are estimated to be about 39 percent. An additional 30 percent of the nitrogen applied to cropland is lost by leaching and denitrification.

Anaerobic fermentation of manures can produce gases with ratios up to 65 percent methane to 35 percent carbon dioxide, at quantities up to 0.46 cubic meters per kilogram of solids, for cow manure (Jewell, 1977) or swine manure (Lapp et al., 1974). A typical beef animal on feed voids about 2.25 kilograms per day of dry weight of recoverable waste. A 1,000-head feedlot would produce 2.25 metric tons per day which could produce 1,035 meters of gas per day with 25,000 million joules of energy (7,000 kilowatt-hours). This would be a maximum level of output and would be decreased by any need for external heat, pumping energy, and combustion losses. Lapp et al. (1975) operated a half-ton pickup truck on compressed feedlot waste methane at 0.19 cubic meters per kilometer. Thus, methane from the manure of one steer could drive a pickup truck 5 kilometers per day. Since many livestock systems are very large in scale (50,000-head feedlots), energy generation from manure might have some merit; and, properly handled, the nutrients remain to be applied back on the land. Nationwide, manures could produce 650 trillion joules of methane gas.

Raising biological material as a direct energy source for combustion has been considered in detail by Roller et al. (1975). The dry-matter yield, cost, and ratio of energy output to energy input for six crops are shown in Table 5. Certainly, it can be seen that, considering equipment, seed, labor, chemicals, feed, and fertilizer inputs, these crops can be used as energy amplifiers. The conclusion reached by Roller et al., however, in considering a number of alternatives for production of food, feed, and fiber, was that economics dictates against diversion of agricultural lands from food production to energy production at this time.

Some authors have portrayed the American farmer as an extravagant spender of energy resources when compared with his counterparts in agriculturally less-developed areas. Generally these comparisons have been made on the basis of use of petroleum fuels. Even including the energy efficiencies of livestock feeding, Nelson et al. (1975) have determined that the American farmer is producing 16 units of energy in plants as food, feed, and fiber, and of these, 1 unit provides food for the American consumer and a second unit is exported, for an energy investment of 1.4 units of fossil fuel. If all energy inputs are considered for food production in the less developed countries, including human and animal labor and the burning of wood, dung, or coal, the picture is quite different. Revelle (1976) determined that the total energy utilized in rural India was 46×10^{17} joules per year compared with 44.6×10^{17} joules per year for American farmers. Makhijani and Poole (1975) similarly found a high energy investment for agricultural production in China, Tanzania, Nigeria, Mexico, and Bolivia, as well as India.

For example, if we consider the energy conversion efficiency of a horse, we find that he is about 25 percent efficient in converting feed energy—grain and hay—into physical work (Morrison, 1949). This is, in fact, comparable to the thermal efficiency of an internal combustion engine. But since we cannot turn off the ignition on a horse at the end of the day, his

TABLE 5. Dry-matter yield, cost, and energy yield for six crops.

Crop	Tons per Hectare	Dollars per Ton	Energy Output per Energy Input
Alfalfa	12.1	41	14
Corn	19.3	29	12
Kenaf	19.5	28	13
Napier Grass	50.5	18	13
Slack Pine	14.5	32	25
Sycamore	16.2	31	17

Source: Roller et al. (1975). Energy inputs include equipment, seed, labor, chemicals, fuel, and transportation to the collection site.

daily work efficiency is about 9 percent. And because he is not worked 365 days per year, his yearly conversion efficiency is on the order of 2 to 3 percent. Therefore, on the basis of combustion efficiency, it would be more efficient to burn hay and grain in an external combustion engine to extract work than to utilize it through a horse.

Similarly, we can estimate the value of human labor. A man can maintain a continuous work level of about 0.75 watts (Suggs and Splinter, 1961a, 1961b) if he is in reasonably good physical condition. If we take the average fuel efficiency of tractors tested at the Nebraska Tractor Test Laboratory (2.6 kilowatt-hours per liter of diesel fuel), then for a 10-hour day, all of the physical work one man can do in a task such as shoveling grain or tilling the soil can be done by a tractor at a fuel expenditure of 0.3 liters. At today's market price of 12 cents per liter (45 cents per gallon) for diesel fuel, the equivalent fuel cost is 4 cents.

In the hot, humid tropics the work output for both man and animal will be considerably depressed. A working animal with a mechanical efficiency of 25 percent must dissipate the remaining 75 percent of the work energy, generally through sweating, radiation, evaporation from lungs or tongue, conduction, or convection. The hot, humid conditions of the tropics reduce the heat dissipation capacity of the animal, forcing an increase in body temperature and total physiological stress. Thus, we can appreciate the adaptation of workers in many tropical areas where a siesta is taken during the peak temperature stress periods.

CONCLUSIONS

The energy balance in American agriculture has been discussed, outlining some of the areas of high energy investment and some approaches to reducing energy investment. Some of the alternative uses of energy resources in the agricultural system have also been discussed. To place the entire picture in perspective, it should be appreciated that the American farmer, by investing 1 percent of our nation's total energy budget in direct energy consumption and 1.9 percent indirectly through chemicals and manufactured inputs, feeds himself and 50 others. It should be recognized that the major energy consumptive areas in the food chain lie in food processing (4.8 percent), in commercial food services such as restaurants (2.8 percent), and in home food preparation (4.3 percent).

The American farmer has received a great deal of attention for his energy consumption. Economic considerations—higher fuel costs—will force him to study the problem and adopt energy conservation practices. Even a 50 percent reduction in on-farm fuel consumption would hardly be detectable in the total energy structure. However, the critical importance of this energy investment for our food resources should be recognized, especially the inelasticity relative to time of input. As energy-use strategies are developed, we must protect the necessary energy resource base of the only major segment of the economy which, through photosynthesis, can yield a net energy gain.

References

Axthelm, D. D. 1962. Report on pumping plants. Department of Agricultural Engineering, University of Nebraska, Lincoln.

Batty, J. C., S. N. Hamad, and J. Keller. 1975. Energy inputs to irrigation. *Journal of the Irrigation and Drainage Division,* ASCE, 101(1R4):293–307.

Bashford, L. L., D. P. Shelton, D. L. Anderson, D. E. Rolofson, and N. W. Sullivan. 1977. Energy and U.S. agriculture: A statewide energy conservation program on Nebraska farms. Report to Federal Energy Administration, Office of Conservation, Washington, D.C. Contract CA-04-60563-00.

Booz, Allen, and Hamilton, Inc. 1976. *Energy use in the food system.* Office of Industrial Programs, Federal Energy Administration. FEA/76/083–D. U.S. Government Printing Office, Washington, D.C.

Brier, B. 1977. Fuel for the farm. *Petroleum Independent* Jan.–Feb.:4–8.

Doering, O. C., T. J. Considine, and C. E. Harling. 1976. Accounting for tillage equipment and other machinery in energy analysis. Energy and Agriculture Conference, Washington University, St. Louis, June 1976.

Economic Research Service. 1974. Energy use in agriculture. U.S. summary by commodity. U.S. Department of Agriculture, Washington, D.C.

Fischbach, P. E., and B. R. Somerhalder. 1971. Efficiencies of an automated surface irrigation system with and without a runoff re-use system. *ASAE Trans.* 14(4):717–719.

Fischbach, P. E., J. J. Sulek, and D. D. Axthelm. 1967. Your pumping plant may be using too much fuel. *Nebr. Univ. Coll. Agr. Home Econ. Quarterly* 19(4):18–20.

Federal Energy Administration. 1977. A guide to energy savings for the field crops producer. June 1977. FEA–USDA report, U.S. Department of Agriculture, Washington, D.C.

Gilley, J. R., and D. G. Watts. 1977. Possible energy savings in irrigation. *Journal of the Irrigation and Drainage Division,* ASCE, 103(1R4):445–457.

Green, M. B., and A. McCulloch. 1976. Energy considerations in the use of herbicides. *Journal of the Science of Food and Agriculture* 27(2):95–100.

Heichel, G. H. 1976. Agricultural production and energy resources. *American Scientist* 64:64–72.

Hoeft, R. G., and J. C. Siemens. 1975. Energy consumption and return from adding nitrogen to corn. *Illinois Research* 17(1):10–11.

Jewell, W. J. 1977. Natural gas from agricultural wastes. *Engineering: Cornell Quarterly* 12:14–24.

Lane, D. E., P. E. Fischbach, and N. C. Teter. 1973. Energy uses in Nebraska agriculture. Department of Agricultural Engineering, University of Nebraska, Lincoln. No. CC255.

Lapp, H. M., D. D. Schulte, and L. C. Buchanan. 1974. Methane gas production from animal wastes. Can. Dep. Agr. Publ. 1528.

Lapp, H. M., D. D. Schulte, A. B. Sparling, and L. C. Buchanan. 1975. Methane production from animal wastes: Fundamental considerations. *Can. Agr. Eng.* 17:97–102.

Makhijani, Arjun, and Alan Poole. 1975. *Energy and agriculture in the Third World*. Ballinger Publishing Co., Cambridge, Mass.

Morrison, F. B. 1949. *Feeds and feeding*. 21st ed. Morrison Publishing Co., Ithaca, N.Y.

Nelson, L. F., W. C. Burrows, and F. C. Stickler. 1975. Recognizing productive energy efficient agriculture in the complex U.S. food system. ASAE Paper 75–7505. American Society of Agricultural Engineers, St. Joseph, Mich.

Noffke, M. H., D. D. Axthelm, and H. R. Mulliner. 1975. The Benedict Project report. Blue River Association of Groundwater Conservation Districts, York, Nebraska.

Pimentel, D., L. E. Hurd, A. C. Bellotti, M. J. Forster, I. N. Oka, O. D. Sholes, and R. J. Whitman. 1973. Food production and the energy crisis. *Science* 182:443–449.

Revelle, R. 1976. Energy use in rural India. *Science* 192:969–975.

Roberts, E. B., and R. M. Hagan. 1975. Energy requirements of alternatives in water supply, use, and conservation: A preliminary report. Calif. Water Res. Cent. No. 155. University of California, Davis.

Roller, W. L., H. M. Keener, R. D. Kline, H. J. Mederski, and R. B. Curry. 1975. Grown organic matter as a fuel raw material resource. NASA Contractor Report 2608. Ohio Agricultural Research and Development Center, Wooster, Ohio.

Schleusener, P. E., and J. J. Sulek. 1959. Criteria for appraising the performance of irrigation pumping plants. *Agr. Eng.* 40(9):550–551.

Splinter, W. E. 1976. Center-pivot irrigation. *Scientific American* 234:90–99.

Steinhart, J. S., and C. E. Steinhart. 1974. Energy use in the U.S. food system. *Science* 184:307–316.

Suggs, C. W., and W. E. Splinter. 1961a. Effect of environment on the allowable work load of man. *ASAE Trans.* 4(1):48–51.

———. 1961b. Some physiological responses of man to workload and environment. *J. Appl. Physiology* 16:413–420.

Van Dyne, D. L., and C. B. Gilbertson. 1978. Estimated livestock and poultry manure and nutrient production in the United States. Economic Research Service. U.S. Dep. Agr. Stat. Bull. January 1978.

World oil crunch seen in the 1980's. 1977. *Oil and Gas Journal* 38:38–39.

6 Management of renewable materials supply

Eric L. Ellwood

It is essential that effective policies and systems be developed to extend the availability of large-scale energy supplies from nonrenewable fossil fuels and to ensure continued production of essential materials derived from nonrenewable resources of limited supply. The options available are the following ones: (1) substitute alternative sources of power, such as nuclear or solar energy, coal, and renewable resources (especially biomass materials), (2) substitute materials from renewable resources, particularly for those derived from fossil fuels, (3) reduce energy use, especially in transportation systems and manufacturing, particularly of high-tonnage materials, and (4) reduce energy consumption in homes and other buildings.

Although the impact of energy shortages is generally recognized, the value of materials and their interdependence with energy is not nearly so well understood. As was stated in the National Academy of Sciences report, *Materials and Man's Needs* (1974), "materials rank with energy and knowledge as basic resources of mankind."

The purpose of this chapter is to point out the potential contributions of renewable resources in relieving the materials supply and energy problems that are developing in the United States and indeed the world. Recommendations are also made which would strengthen the opportunities of the United States to use its renewable resources and to develop the knowledge to assess the best and most appropriate use of renewable resources in the national interest. The presentation draws substantially on a report of the Committee on Renewable Resources for Industrial Materials (CORRIM, 1976), of which the author was a member.

DEFINITION OF "RENEWABLE"

Renewable resources, as discussed in this chapter, are of biological origin —primarily vegetation—and are basically the products of bioconversion through the driving force of solar energy. Two categories of renewable resources may be identified: primary renewable resources are wood and its extractives, certain agricultural crops, cotton and other fiber-producing crops, leather, and seaweed; secondary renewable resources include the wastes and residues generated by the use of vegetation and animals. The second category includes residues from the harvesting and processing of wood and agricultural crops and from animal raising and processing. (Sand- and clay-based materials might also be considered renewable, but they are not a central subject of this discussion.)

OVERVIEW OF MATERIALS USE

An overview of the consumption of principal high-tonnage materials is found in Table 1, which shows that renewable organic materials, primarily wood, constitute 6.6 percent of the total.

Nonrenewable resources—of which oil and natural gas make up about 30 percent—amount to 93.4 percent of the total. Approximately 6 percent of the oil and natural gas is converted into chemicals and plastics; the balance is used for energy. Currently the United States imports more than 40 percent of its oil at prices about four times higher than four years ago.

Although currently (1976 figures) the total annual tonnage of wood produced in the United States (237 million tons, see Table 2) is more than the combined tonnage of steel, cement, plastics, and aluminum, an additional 782 million tons of lignocellulosic residues from wood, agricultural crops, and animal processing are potentially available (see Table 3). The largest tonnages of these secondary renewable resources are cow manure, urban and municipal solid waste, and agricultural residues.

Many of us whose careers have been in wood-utilization research and development are convinced that the full potential of wood as a useful resource has nowhere near been reached. Perhaps one reason is that wood has been abundant in the United States for such a long time and is seen as a

TABLE 1. Use of basic raw materials in the United States.

Raw Material	Total Weight (percent)
Nonrenewable Resources	
Sand and Gravel	21.2
Stone	20.0
Petroleum	18.3
Natural Gas	11.8
Coal	11.8
Other Nonmetallic Materials	7.2
Metals	3.1
Total	93.4
Renewable Resources	
Forest Products	6.34
Cotton	0.05
Other Fiber	0.06
Seed Oil	0.07
Animal Fats	0.06
Natural Rubber	0.03
Total	6.60

Source: Committee on Renewable Resources for Industrial Materials (1976), p. 38. Based on the Report of the National Commission on Materials Policy (June 1973).

material with uses more appropriate to earlier times. Until recently, oil and natural gas have been cheap and readily available, and their use in the development of petrochemicals and the associated plastics industries has also contributed to an almost complete collapse of research and development in the chemical utilization of wood, agricultural chemicals, and coal. Thus, although wood is a viable material for the production of structural, decorative, and printing and packaging materials, its per capita use, along with that of other natural products such as cotton and wool, has steadily given way to the use of products derived primarily from oil and natural gas. Also, the use of light metals and ceramic materials has increased.

The agricultural residues, because of their dispersed distribution, pose problems of collection as well as of seasonal availability and of alternate use as fertilizers. Urban and municipal solid waste, of which two-thirds is organic, poses problems of economic viability because of separation difficulties and the scale of these operations.

TABLE 2. Approximate annual use of lignocellulosic renewable resources in the United States.

Material	Tons
Hardwood	90,384,000
Softwood	146,751,000
Total Wood	237,135,000
Cotton Lint	2,541,000
Cotton Linters	344,000
Total Cotton	2,885,000
Other Fiber Crops (flax, ramie, hemp, sisal, kenaf, etc.)	140,000
TOTAL	240,160,000

Source: CORRIM (1976).

TABLE 3. Lignocellulosic residues and by-products.

Source	Tons
Wood Residues	
Logging	50,000,000
Manufacturing	15,000,000
Total Wood	65,000,000
Urban and Municipal Solid Waste	160,000,000
Cow Manure	210,000,000
Cereal Straws	132,000,000
Cornstalks	125,000,000
Miscellaneous	90,000,000
TOTAL	782,000,000

Source: CORRIM (1976).

ENERGY PRODUCTION FROM RENEWABLE RESOURCES

Approximately 75 percent of energy sources in the United States are derived from oil and natural gas. Energy from coal ranks next at 20.4 percent (see Table 4). Energy production from nuclear power is still on a small scale. Although substantial expansion of power production from coal is appropriate at this time, problems remain concerning the environmental impact of mining and the control of sulphur emissions in coal-burning operations. The use of lignocellulosic residues for direct energy production by combustion has much less of an environmental impact than coal mining, and toxic emissions are much less of a problem. Use of wood for energy production can make a substantial contribution, but it cannot completely compensate for the reduced availability of fossil fuels. If the total annual tonnage of all lignocellulosic residues and by-products were economically collectible for energy production by direct combustion, the energy contribution from this source would be about 12 percent of the nation's energy needs. The best opportunities for use of these residues for energy production would appear to be through direct combustion at major wood-processing plants where residues are already collected and concentrated. In the South the pulp and paper industry—one of the major industrial consumers of energy—now supplies up to 45 percent of its own energy from process wastes. Additional opportunities may exist in certain regions to use low-grade biomass and wood residues to provide process energy for small- to medium-size industries now dependent on oil, natural gas, or propane. The idea of growing "energy plantations" specifically to provide fuel for large-scale thermal power plants may offer possibilities for countries that have an excess of inexpensive wood. This approach seems less practical for the United States, with its large coal reserves.

The use of domestic firewood in conjunction with improved home-combustion systems and fireplaces could substantially reduce the power required for winter heating. The use of urban solid waste for energy production by direct combustion is undergoing large-scale evaluation in several locations, such as St. Louis, Baltimore, Hempstead (Long Island), and Boston. However, no country as yet has found power production from urban residues to be economical. Development of effective systems to sort

TABLE 4. Energy sources in the United States (1970).

Source	1.05×10^{18} Joules	Percent
Oil and Natural Gas	52.0	75.7
Coal	14.0	20.4
Hydroelectric	1.0	1.5
Nuclear (1974)	0.7	1.0
Wood Residues		
(pulp and paper industry)	0.8	1.1
Fuelwood	0.2	0.3
Total	68.7	100.0

Source: CORRIM (1976).

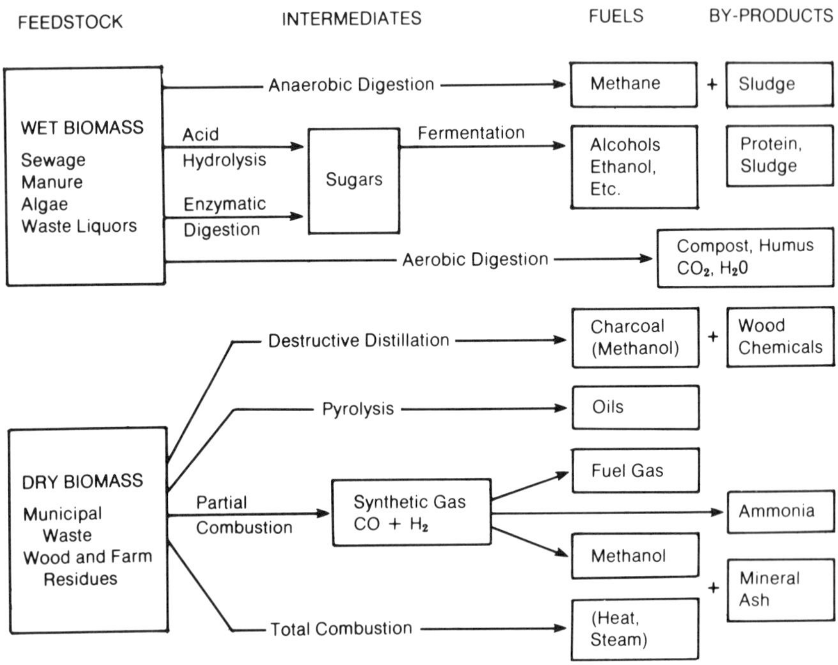

FIGURE 1
Biomass to fuel conversion routes.

out materials for recycling from solid waste and the development of improved combustion units are important keys to the effective use of this material for more than just landfill.

Besides direct combustion, which appears to be the most attractive alternative for power production, energy may be produced from lignocellulosic materials in a variety of ways (see Figure 1). Wood gasification, pyrolysis, and technologies for producing methane and methanol call for more extensive examination, particularly as fossil fuel costs rise.

OTHER USES FOR RENEWABLE RESOURCES

Structural Uses

The chief use of wood (60 percent) is for lumber and plywood in building and construction. This use takes advantage of the natural properties of the material, preserving the integrity of the structure of the wood. In addition, a number of reconstituted wood products made from veneers, chips, fragments, and fibers have been developed in the last two decades. Because of the diminishing availability of large trees and the increasing economic need to use smaller trees and residues, it is expected that the use of reconstituted wood products will increase at the expense of solid wood (lumber). Figure 2 illustrates the variety of products generated by a lumber manufacturing operation.

What is important here, however, is that the manufacture and use of structural wood components require less energy than is used for structural or building components manufactured from nonrenewable resources (see Tables 5 and 6). It would appear that an overall twofold advantage may be gained by using renewable resource materials, such as wood, rather than using, say, a nonrenewable metal in certain structures: (1) conservation of an essential nonrenewable material, and (2) energy saving through the use of a less energy-intensive material.

In order to assess a materials-substitution strategy or a new technology, the CORRIM study applied a Reference Materials System (RMS) which

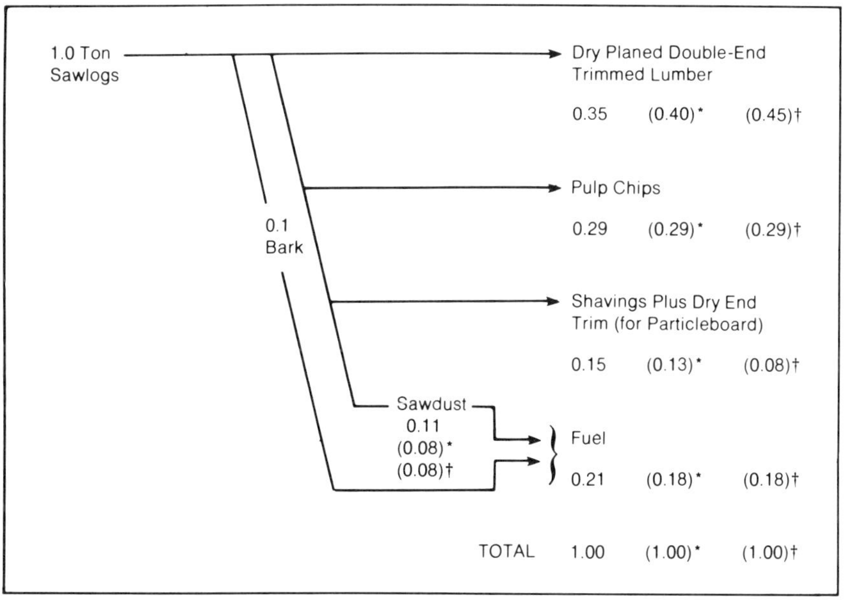

FIGURE 2
Materials flow for softwood lumber (products as percentage of one oven-dry ton).
*Predicted product and by-product recovery, 1985.
†Predicted product and by-product recovery, 2000.

Source: CORRIM (1976), p. 140.

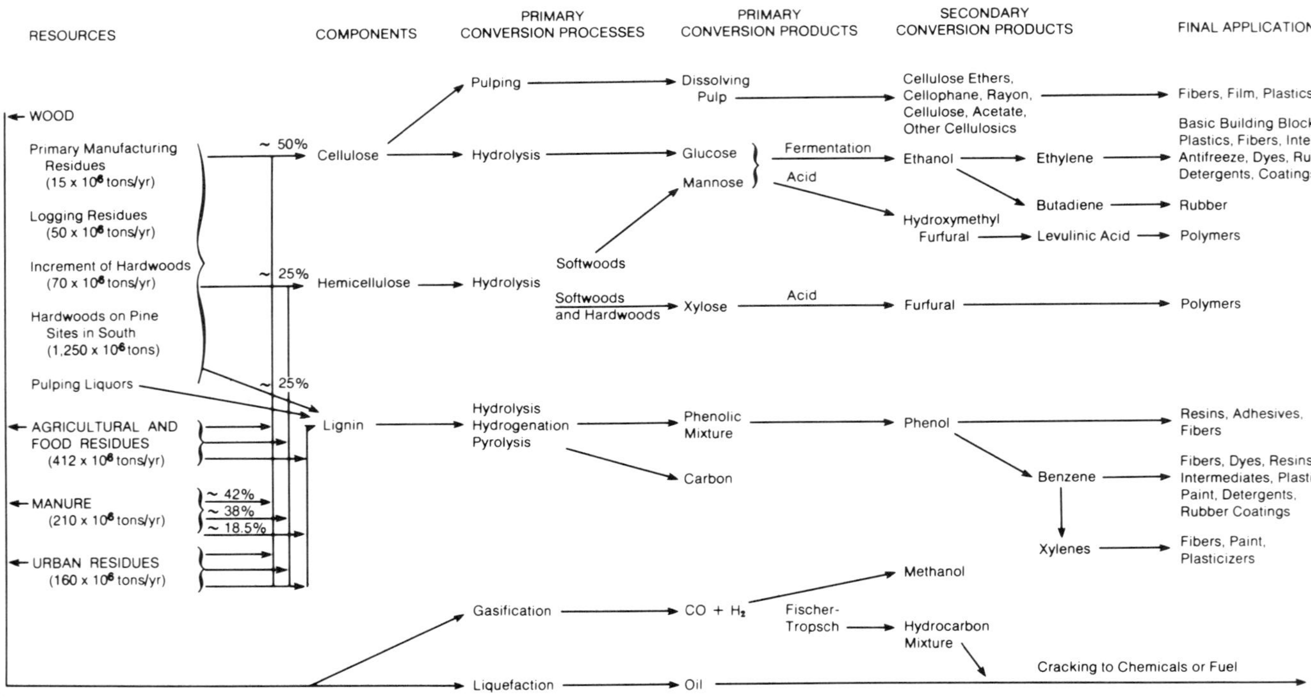

FIGURE 3
Potential flow of wood and agricultural, municipal, and feedlot residues for chemical use.
In the chart above, 90.6×10^6 tons of cellulose will produce 48.9×10^9 pounds of ethanol, which can be converted to 23.52×10^9 pounds of ethylene and 3.66×10^9 pounds of butadiene. These latter two figures were the actual 1974 U.S. production derived from petroleum sources. Also, 25.6×10^6 tons of lignin will produce 17.92×10^9 pounds of phenol, which can be converted to 11.07×10^9 pounds of benzene (the 1974 annual production).

TABLE 5. Energy requirements to produce commodities (1.05 x 10^9 joules per oven-dry ton).

Commodity	Total Energy Requirement	Available Residue Energy	Net Energy Requirement
Wood-Based Commodities			
Softwood Lumber	7.755	8.313	2.909
Oak Flooring	8.741	11.388	3.050
Softwood Plywood	9.699	3.697	6.002
Medium Density Fiberboard	11.232	2.741	8.491
Underlayment Particleboard	13.916	1.529	12.387
Nonwood-Based Commodities			
Gypsum Board			3.52
Concrete			8.52
Concrete Block			8.77
Clay Brick			9.06
Steel Joists			50.32
Aluminum Siding			200.47

Source: CORRIM (1976).

was developed at the Brookhaven National Laboratory in connection with research supported by the Energy Research and Development Administration (see Figure 3). The system diagrams trajectories from a specific resource to a specific use, and shows the materials and energy used for all activities associated with the manufacture of the product. Thus, depending on the detail supplied, it is possible to ascertain capital costs, manpower costs, energy use, and environmental costs of alternative systems and to evaluate

TABLE 6. Comparative energy requirements for various building construction systems.

System	Siding	Exterior Framing	Interior Framing	Flooring	Rafters and Trusses
Wood Studs		1	1	1	1
Plywood Fiberboard	1				
Steel			8	50	7
Aluminum	5	20	12		
Brick	25				

Source: CORRIM (1976).

perturbations of these systems. The methodology is a most useful way to examine alternatives, although it is apparent that the availability and organization of data for renewable resources require further study.

The several RMS studies comparing building components and systems did not deal with matters of maintenance, insulative effectiveness, or effective life of the structures. For a complete evaluation, these and other factors need examination.

Packaging

Containers and packages consume large amounts of energy. RMS trajectories developed to compare mass and energy values for half-gallon milk containers made from polyethylene and from paper show 40 percent higher energy use for manufacture using the plastic material. Similar trends were found in comparisons of paper and polyethylene bags. On a straight energy conservation basis, renewable resource materials would appear preferable for many packaging applications.

Renewable resource materials (primarily paper and paperboard) still dominate the packaging field. As costs of energy rise, costs to the paper industry should rise at a slower rate because of the industry's high degree of energy self-sufficiency through the use of its own process wastes. Paper-based materials for packaging might therefore be expected in the long run to gain in economic advantage over petroleum-based plastic packages.

Textiles and Fibers

In 1972 the United States consumed 5.9 million tons of fibers. Some 52 percent of these fibers came from nonrenewable materials (47 percent petroleum based, 5 percent glass), and 48 percent came from renewable resources (34 percent cotton, 1 percent wool, and 13 percent rayon and cotton linters).

Synthetic fibers in textiles came to be widely used because they were cheaper than natural fibers and because they had certain characteristics that were in demand, such as strength and ease of care. The use of cellulosic fibers, such as rayon derived from wood, has been declining, mainly because of the high capital and energy costs of production. The RMS trajectories developed for cotton, synthetic fibers, and regenerated cellulose fibers show 14.67, 30.61, and 34.71 kilowatt-hours, respectively, as the energy needed per pound to produce, fabricate, and finish these materials (see Table 7). This is one of the few instances in which a natural product (regenerated cellulose) proved to be more energy intensive than an entirely synthetic product.

TABLE 7. Energy consumption for textile fibers (in kilowatt-hours per pound of fiber).

Energy Use	Fiber		
	Cotton	Wood-Based	Noncellulosic
Energy Content of Raw Materials	0.20	1.16	6.28
Fiber Production	3.55	22.09	11.36
Fabrication	4.40	3.93	5.44
Finishing	6.52	7.53	7.53
Total	14.67	34.71	30.61

Source: CORRIM (1976).

Chemicals, Feedstocks, and Plastics

The three main constituents of wood and agricultural residues, which form the bulk of renewable resources, are cellulose, hemicellulose, and lignin. These three materials are the most abundant naturally occurring high polymers on earth. From an entropy standpoint it is advantageous to use them in a "least modified" condition, such as for lumber, plywood, paper, and board, and in reconstituted wood products and possibly regenerated cellulose. But much of the available biomass is not of a quality that can be used to produce the "higher order" products. Therefore, from the standpoint of supply, there is reason to believe that high-tonnage chemicals, feedstocks, and hence plastics can be derived from lignocellulose resources by way of chemical transformation to the molecular level without diminishing the availability of wood for existing uses.

The total U.S. production of plastics, noncellulosic fibers, and synthetic rubber amounts to more than 18 million tons. This production consumes about 6 percent of the nation's available oil and natural gas as a raw material. The amount of wood needed to substitute completely as a raw material for the production of the 18 million tons of chemicals and plastics approximates 60 percent of that now used by the pulp and paper industry. It is believed that about 95 percent of these materials can be derived from cellulose, lignin, and hemicelluloses. Although technology is known to effect many of the necessary conversions, research and development in this field have been almost static because of the price advantage of chemicals synthesized from petroleum sources. But even now, certain systems based on lignocellulosics appear to be economically feasible.

Cellulose, a long-chain polymer, can be modified to produce rayon and other products without substantial modification of the cellulose. However, the price-property advantage of petrochemically derived plastics has to some extent truncated research and development on modified cellulose derivatives which could otherwise reduce the demand for nonrenewable resources.

With regard to chemical feedstocks, most of the synthetic polymers and plastics produced in the United States can be derived from ethylene (47 percent), phenol (36 percent), and butadiene (12 percent). All of these chemicals can be derived from lignocellulosic materials essentially breaking down the wood polymers, as shown in Figure 4. Thus ethylene and butadiene can be produced from ethanol, which is a product of the fermentation of glucose obtained from cellulose by hydrolysis. Butadiene is a primary feedstock for synthetic rubber, and ethylene for thermoplastic polymers.

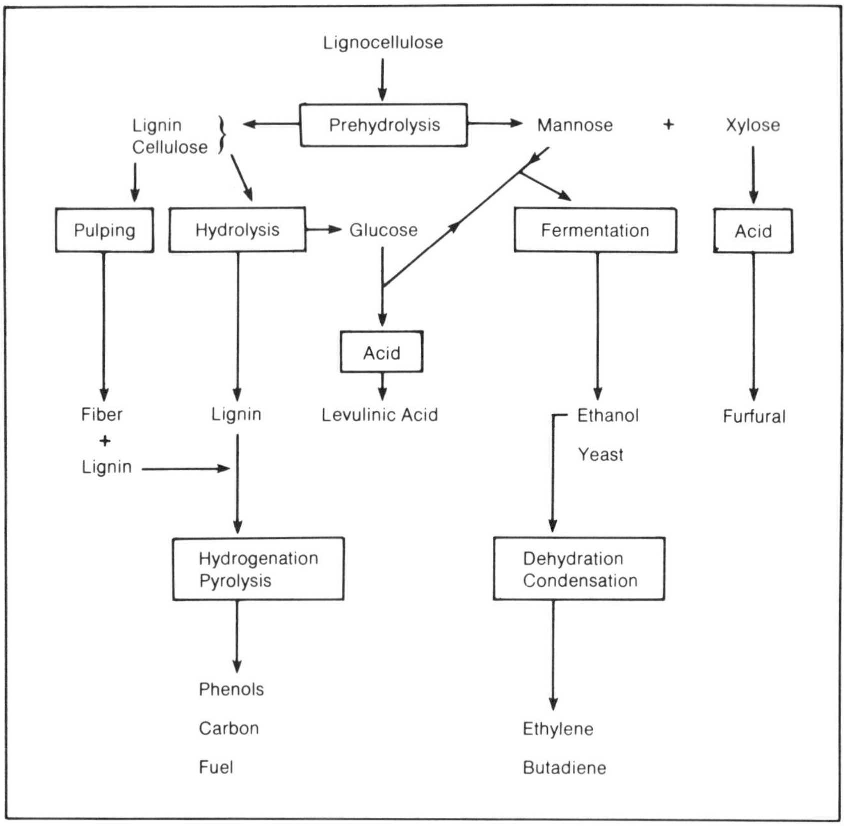

FIGURE 4
Schematic flow chart of a sample lignocellulosic chemical plant.

Lignin, which contains aromatic structures, can be transformed into products such as phenols and benzene. Hemicelluloses, the components of which are pentose and hexoses, can be hydrolyzed into xylan and transformed into xylitol (a noncaries producing sweetener) or furfural, which was at one time used as a feedstock for nylon 66 production.

Prompted by the increasing cost of crude oil, a number of preliminary studies have been made in Europe, Canada, and the United States on the potential of wood as a basis for chemical feedstock production. Several of these studies do indicate the possible economic feasibility of adopting multiproduct or integrated systems rather than producing a single chemical.

In the multiproduct system, yields from 100 pounds of hardwood of ethylene (8 pounds), butadiene (5.6 pounds), coproducts of monomeric phenol (8.3 pounds), and furfural (12.3 pounds) could be realized with a final product value of five to six dollars per 100 pounds of oven-dry wood, using present technology. Wood residues can be obtained at one dollar per 100 pounds (oven dry) delivered to the plant.

Estimated sale values of the several feedstocks studied show about a fivefold increase over the economic value of the original wood. If theoretically possible yields could be obtained through improved technology, then yields and values would be approximately 75 percent more than those shown. Also, the possibilities for chemical use of carbohydrates produced in the waste streams of pulping operations need to be examined. Currently these waste streams are combusted for energy production.

Until 1929 practically all U.S. ethyl alcohol was produced from agricultural crops. But lower-cost synthetic alcohol from petroleum has come to dominate the market.

The use of lignocellulosic materials for chemical and feedstock production does appear promising as a means of conserving energy and materials. Areas in need of technological development can be readily identified: (1) collection, transportation, and sorting technology, (2) new technology for production of chemical or regenerated cellulose and diversification of its properties and performance, (3) conversion of lignin to phenols and to applications based on its inherent properties, (4) conversion of cellulose to sugars to optimize yield, and (5) economics-engineering analysis of wood compared with coal.

Another alternative to chemical production from lignocellulosic materials is food production, and this is being done to a limited extent by using particularly cellulosic and hemicellulosic waste streams either to furnish carbohydrates directly or to produce single-cell proteins. Obviously the production of low-cost sugars from lignocellulosics would create opportunities in this field.

EXTRACTIVES AS FUTURE RAW MATERIALS

With some exceptions, extractives are produced as by-products during the processing of materials for other primary uses—that is, wood, food, and feed. Of the 11.5 million tons produced in 1972, the majority are vegetable oils (70 percent), tallow (22 percent), and wood extractives (6.4 percent). Extractives occupy a diverse and critical place in a variety of industries. Naval stores by-products from pulping of pines, and tallow and grease from the rendering of beef, appear to be the most likely substitutes for petroleum-based materials.

RESEARCH AND ASSOCIATED EDUCATIONAL PROGRAMS

One of the most striking conclusions derived from an overview of the research and development in use of renewable resources is the relative poverty of this field in relation to the materials field generally.

Not only is utilization research on renewable resources conducted at a low level in the United States, but basic research on the nature and properties of the natural organic materials which make up these resources is also minimal. In fact, research in wood science and technology has been in relative decline over the last two decades—at both the federal and the university levels. In the last decade, several universities have phased out their programs in wood science and technology primarily because of lack of support. Currently only eight major universities have significant research programs in wood science and technology.

The U.S. efforts also suffer by comparison with European and Russian research. For example, four times as many research papers on lignin-use research have been produced in the Soviet Union as in the United States in the last ten years. Technological advances in the processing and use of forest resources are more frequently generated in foreign countries than in the United States. The overall effect is a shortage of scientists with special expertise in biological materials.

In the industrial sector, research expenditures have also been at a low level compared with other materials, because wood products are treated essentially as high-volume, low-cost commodities. Consequently, industrial research and development expenditures in the forest products industries have been less than 0.5 percent of the sales value of the products.

In recent years, several leading forest resource industries have begun to expand their research and development expenditures substantially, perhaps in response to the increasing costs of raw materials and energy, and the substantial increases in the capital costs of plant construction. But most of this research is designed to produce existing products more efficiently rather than to initiate new technologies.

There is little question that if the United States is to benefit from the potential inherent in lignocellulosic materials, financial support must be obtained to modernize technology and extend research and development.

RECOMMENDATIONS AND CONCLUSIONS

1. A national materials policy is needed which coordinates and integrates the attempts to minimize depletion of essential materials and to conserve energy. This is necessary to gain time to develop new technologies that can deal with the eventual unavailability of fossil fuels and certain scarce materials.
2. Renewable resources have the potential of reducing the depletion of essential nonrenewable materials and conserving energy by their use. However, a substantial expansion of basic and applied research is needed to modernize conversion technology and to develop new applications based on strengthening the science of the properties and behavior of biomass components.
3. With research and development in natural renewable resources itself a depleted resource, a limited number of "critical mass" university research centers should be established with federal support at institutions with appropriate faculty and facilities to conduct such research and produce scientists with expertise in this field.
4. The Reference Materials System methodology developed by the Brookhaven laboratories provides a valuable starting point in the development of materials-substitution and research strategies. This approach should be further developed.
5. Because of the relative dispersion (compared with oil and coal) of lignocellulosic materials, harvesting and collection technology is vital in expanding the utility of renewable resources. Harvesting and collection systems should be given high priority in research.
6. The potential use of lignocellulosic (biomass) materials for energy production and for production of chemical feedstocks and plastics should receive high-priority research and development support through to the pilot plant stage.
7. Because most of the tonnage of renewable resources is used in building, research is recommended on more efficient means of using wood in structures, with a view to conserving energy, manpower, capital, and materials.
8. Cooperation must be strengthened between government agencies,

universities, research institutions, and the private sector concerned with renewable resources. Such cooperation offers the best possibility for developing new practical technologies.

Knowledge is power. It is certainly in the national interest to develop our knowledge of renewable resources—the least effectively utilized of all our resources.

References

Committee on Renewable Resources for Industrial Materials. 1976. *Renewable resources for industrial materials*. Board on Agriculture and Renewable Resources, Commission on Natural Resources, National Research Council, National Academy of Sciences, Washington, D.C.

National Academy of Sciences. 1974. *Materials and man's needs*. Summary report of the Committee on the Survey of Materials Science and Engineering. Washington, D.C.

7 Public policies affecting land devoted to agriculture and forests

B. Delworth Gardner and John A. Zivnuska

Because it cannot be physically moved to accommodate human preference, and because it has become scarce relative to the demands on it, land is now a very "social" resource. Its use is tightly regulated by public policy of one kind or another in most modern societies.

In earlier times, when land was plentiful and capital was relatively scarce, economic growth could be stimulated by bringing more land into production. Public policy attempted to create incentives for putting land and water to beneficial use through the Homestead and Reclamation Acts and other legislative enactments. Land use was predominantly single-purpose, and conflicts between uses were minimal.

As population and economic growth continued, however, competition for land increased and conflict became inevitable. Instead of encouraging growth and land utilization, public policy in land use often became exclusionary and growth-inhibiting. Instead of granting essentially free land, policy took the form of acreage allotments, exclusionary zoning, subdivision control, building codes, use easements, heavy land-use taxes, and so forth. Land has become so scarce that many groups are fearful that their uses will be choked off completely. They are opting for single-use legislation and land-reservation policies that will use the police power of the state to guarantee only their access to land. The land market as an allocating mechanism is being steadily eroded, and the allocating function is being controlled by government regulators and planners.

This chapter attempts to identify and partly evaluate those public policies that influence the allocation of land to agriculture and forests. The first section discusses some economic "first principles" that provide the rationale for market allocation of land and the justification for societal interference in the land market. The second and third sections then deal with public policies affecting land allocation to agriculture and forests.

THE THEORY OF MARKET ALLOCATION

A simple idealized economic model can illustrate how the market works to allocate land and establish land values. Assume that putting land parcel A to given use X (say, agricultural production) rules out using it for any other simultaneous purpose. (This assumption will be relaxed later.) Suppose, however, that if parcel A were not used in X, it could be used in Y (say, as developed urban land). Now suppose A is combined with other inputs in a production process that yields a given quantity of output. For simplicity, suppose that perfectly competitive markets exist for these inputs and for X. This means that the user of A is a price taker in both input and output markets. The term "user surplus" will be used to represent the net value of putting A to a given use for a fixed time period (say, one year). Assuming that all benefits and costs of using A in X can be quantified in monetary terms and that benefits are fully captured by the landowner and costs are fully incurred by him, user surplus will equal annual dollar benefits generated by the sales of X minus the dollar nonland costs of putting A to this use. The optimal amount of X produced will be where the incremental costs of producing X are equal to the price of a unit of X.

Provided that A yields a larger annual user surplus at the optimal level of X than a corresponding optimal level of Y, that the annual user surplus can be captured in perpetuity, and that there is a perfectly competitive land market, the market price for A will be the user surplus divided by the relevant interest rate. A number of potentially thorny issues are neglected by making these assumptions, but their resolution cannot be accomplished in this space. In any case, under the assumptions specified, the market will allocate parcel A to use X; and this allocation will be socially efficient in the sense that no other allocation will generate a greater user surplus and thus a greater aggregate net return.

It follows that if all services having value emanating from land were fully and completely expropriable by landowners (and only by them), and if all costs of utilization were borne by them, the competitive market would ensure that land would be allocated to its socially most productive use or uses. Under this system, those who are able and willing to pay most for land will acquire it and apportion it to various uses. The resulting allocation would maximize the aggregate (social) value of land.

The rub in this argument is that often the costs and benefits of managing land in a given use are not fully imposed on and captured by the landowner. There may be indirect costs and benefits that fall on other individuals. Economists call these "externalities." Where they occur, a disparity exists between the *private* valuation of land by its owner and the *social* value of land to all those who are affected by its use. The competitive market price reflects only the private valuation, and obviously it may understate or overstate the social value.

Even in the absence of externalities, the community may not utilize the market for allocation decisions. Such is the case with so-called public goods. These goods have certain characteristics that distinguish them from private goods. They tend to be indivisible in character; thus, an individual consumer cannot vote in the market for how much of the public goods he wants. A second characteristic is the physical impossibility or costliness of controlling access to consumption of these goods. This implies that users cannot be charged. Finally, the amount of these goods taken by one consumer does not ordinarily affect the quantity available to other consumers. One or more of these conditions are often encountered in certain uses of land such as public parks, open space, wilderness, and watersheds.

Although the market will not allocate resources optimally where externalities and other forms of market failure exist, it should not be assumed that the political process will always do a more efficient job than the admittedly imperfect market. The political sector will generate, at best, a politically efficient allocation. Whether this allocation will also be economically efficient depends on the seriousness of market failure and how well the political decision makers consider all the relevant economic interests of various land users.

LAND UTILIZATION IN AGRICULTURE

Some data on agricultural land use and conversion will serve as a useful reference point (Cotner, 1976). The cropland base in the United States is approximately 385 million acres (156 million hectares). About 1.3 million new acres (526,100 hectares) are added to this base annually via land clearing, irrigation, and drainage developments, and by cropping land previously held out of production. On the other hand, about 2.7 million acres (1.1 million hectares) of cropland are lost to agricultural use each year.

Thus, there is a net loss of 1.4 million acres (566,600 hectares) or about one-third of 1 percent of the national base. But even this very small figure may overstate the seriousness of the loss nationally. The Soil Conservation Service has estimated that there are 266 million acres (108 million hectares), or 69 percent of the present base, of potentially tillable land in pasture, forest, and other uses that could be added to the cropland base if it were economically feasible or deemed socially necessary. One view, therefore, is that the land base is quite adequate to provide for agricultural needs in the United States for the rest of the 20th century at least (essentially the view held by Krause and Hair, 1975).

History tends to support this optimistic view. The total land area devoted to crops in the United States has been reasonably constant since World War II. The 1949 cropland acreage was 387 million (156.6 million hectares)—a 50-year high. By 1969 this figure had dropped to 333 million acres (134.8 million hectares), but it rose again to an estimated 367 million acres (148.5 million hectares) in 1976. Throughout much of the 1950's and 1960's the prices of most basic farm commodities in the United States were supported above world market levels. Various land set-aside policies were invoked, ostensibly to limit supply in order to support prices at pegged levels. From 37 million to 65 million acres (15 to 26.3 million hectares) were kept out of production during the years 1961–72. This acreage was 14 to 24 times the annual loss of cropland to other uses. By 1976, however, most of this set-aside acreage had been returned to crop production. The reasons for this are indicated in the discussion which follows.

These data are by no means the whole picture. In a market economy, an unfailing signal of increasing scarcity is rising relative prices. It is relevant, therefore, to inquire about the movement in agricultural land prices.

The *real* values per acre for agricultural land (in 1972 dollars) and for that land plus the improvements for the post–World War II years have nearly tripled over the period. The tremendous profitability of holding agricultural land for capital gains through this period is shown in Table 1. The numbers in columns 3 and 4 indicate the average real rates of return from holding land for the periods specified. For example, if a typical parcel of agricultural land were purchased in 1946, held until 1975, and then sold, the average real rate of return on the original land price would have been 3.25 percent per year. Unfortunately, no data are available on the variance of these returns, either cross-sectionally or over time. It should be emphasized that these are returns corrected for inflation. As such, they are very high compared with real returns from holding other capital assets. Real returns to capital in the economy as a whole have averaged just under 2 percent in this century (Tatom, 1976).

Two other points are worth mentioning. The first is that the returns to

TABLE 1. Average real rates of return yielded by holding agricultural land, 1946–1975.

Period	Number of Years	Land Only (percent)	Land and Improvements (percent)
1946–1975	30	3.25	2.84
1951–1975	25	3.73	3.09
1956–1975	20	4.03	3.42
1961–1975	15	3.84	3.37
1966–1975	10	3.52	3.13
1971–1975	5	6.13	6.00

land only are consistently higher than the returns to land plus improvements, regardless of the period considered. This corroborates the point made earlier that the returns to land are higher than the real returns to capital in the economy as a whole. The other significant point is the tremendous increase in the rate of return in the last five-year period, 1971–75.

The steady post–World War II rise in the real value of agricultural land is one of the most puzzling phenomena of our times. Numerous explanations have been offered, and some of them are indeed plausible.[1] A preponderance of scholarly opinion, however, would argue that the recent rapid increase in real farmland prices is not attributable wholly to comparable increases in user surplus derived from agricultural production. Still, it is obvious that the profitability of agriculture will continue to play a crucial role in agriculture's ability to compete for land.

Agricultural Price, Income, and Trade Policies

Any discussion of factors explaining the acreage devoted to agricultural production should begin with national agricultural price, income, and trade policies, since their impact on cropland acreage is considerable. There are direct and indirect effects. Direct effects come in the form of land set-aside programs needed to maintain support prices or income levels. Indirect effects encompass a broader set of issues affecting user surplus and the profitability of putting land to agricultural uses.

It has already been pointed out how the high price-support policies of the 1950's and 1960's required large areas of land to be removed from crop production. Although some earlier acts started the movement away from the high, rigid price-support policies of the 1940's, 1950's, and 1960's, it was the Agriculture and Consumer Protection Act of 1973 that produced the most abrupt shift in price and income policy. Under this act, market prices were permitted to seek equilibrium levels, and farmer incomes would be supported by a deficiency payment reflecting the difference between the market price and an announced target price. But the deficiency payment could be paid only on the production from acreage under allotment.

Quite fortuitously, several circumstances occurred on the world scene in 1973, 1974, and 1975 to drive market prices beyond announced target prices. The devaluation of an overvalued dollar and crop failures in several places around the world were the most critical factors. Exports of American crops have exceeded $20 billion in each of the last four years, and the value of exports has increased now for seven successive years. It has been estimated, in fact, that the crops from one-third of the total U.S. cropland were sold abroad in 1976. It is this rise in world food prices above domestic market prices that has made it profitable to produce and export agricultural commodities and has pulled idle U.S. cropland back into production.

Under the new 1977 act, target prices for wheat, rice, cotton, and the feed grains have been increased substantially over previous levels. Stocks of these commodities in both private and government storage have increased sharply the past two years and promise to increase further in the months and years ahead. In an attempt to control aggregate supplies and reduce deficiency payments to producers, the Carter administration has invoked new acreage set-asides for wheat (20 percent) and the feed grains (10 percent). Early enrollment figures, however, indicate that the majority of producers are electing to remain outside the programs so that they might be free to plant whatever acreages they desire. Unless this situation changes, the set-asides may not have a substantial impact on the total acreage in crops.

Obviously, an agricultural price and income policy that permits domestic market prices to be below world prices is a crucial factor in determining the acreage that will be devoted to agricultural crops. It must be kept in mind that there are still many restrictions on the free movement of agricultural goods. Tariffs, quotas, and embargoes are commonplace. Negotiating teams are striving to reduce trade barriers, but progress has been very limted. One thing is fairly clear. If there are major changes in trade policy affecting the competitive position of American agricultural products, the impact on the acreage committed to cropland in the United States will be considerable.

Direct Land-Use Controls and Taxation

Since the early 1960's, many land-use controls have been created in response to (1) the demand for open space by the predominantly urban population, (2) the costs of suburban sprawl, (3) the pressures of population growth in the world, (4) the apprehensions about whether there is sufficient agricultural land, (5) an aroused ecological awareness, and (6) concerns about increased farmland tax burdens. A number of diverse measures to control land use have been initiated at the federal, state, and local levels.

One method of controlling land use offers full or partial compensation for the financial losses suffered by landowners because of the existence of controls. One such tool which shows some promise of success in preserving agricultural land is the so-called Transfer of Development Rights (TDR). Under this system, some local governmental unit identifies the degree of development that is to be permitted. The developer may then purchase development rights in designated preservation zones from landowners who are willing to forgo development in order to sell the rights attached to their land.

The more traditional land-use controls within cities, towns, or suburbs are noncompensable zoning restrictions. California's Land Conservation Act is an example of zoning for aesthetic, recreational, or resource management and can be used to preserve agricultural land.

This question of reserving prime land for agricultural use has received much attention lately. It is not an easy problem to solve. For nearly two decades, the average annual loss of agricultural cropland to urban development has been about 500,000 acres (about 202,000 hectares). This may seem to be a large amount, and in some areas these losses include the most productive land. But when viewed in the context of the current national cropland base, this loss is only 0.13 percent and is therefore relatively insignificant. Other factors account for a much larger change in the cropland base than encroaching urbanization. Furthermore, this loss certainly has little to do with the world food problem and our ability to feed the hungry poor.

This is not to say that losses to urban development are not significant to local communities, and to the ability of agriculture to remain a viable industry in an area. The community may have many reasons to keep agriculture profitable and the land in agricultural use. Most of the reasons will be recognized as examples of externalities, as discussed in the first section. Some of these reasons are (1) to preserve open space for aesthetic and recreational purposes, (2) to prevent urban sprawl and increased costs of providing urban services and public utilities, and (3) to shift the demand for urban development to unused parcels of already developed urban land and thus to a greater intensification of land use.

Most discussions of the prime-land question tend to focus on alleged social benefits. They often neglect the economic costs of most reservation policies. An example will make the point. Suppose that land parcel A has an annual user surplus of $100 per acre in urban development and $50 in agriculture. An alternative parcel B has a user surplus of $150 per acre in urban development and $90 in agriculture. It is clear that the unfettered land market will select B for development if there is sufficient urban demand for only one of the competing parcels. It is also clear that this is socially optimal, disregarding all externalities. Society forgoes $90 in agriculture to get $150 in urban use or captures a rent of $60 by the shift. Alternatively, if A had been chosen, society would get a rent of only $50. It is obvious that the appropriate rule for efficient allocation is to take for development the parcel with the most rent.

The trouble with most prime-land reservation proposals is that they look to agricultural productivity criteria exclusively in deciding which parcels to keep in agriculture. (These criteria include soil depth and texture, gradient, drainage, water availability, and so forth.) Parcel B would almost surely have been selected for agriculture in our example, since the user surplus in agriculture is greater than for A. The correct criterion of maximum rent would have allocated parcel B to urban use and parcel A to agriculture rather than the reverse.

Taxation policies also have a direct effect on land use. There are three major classifications of taxation, each with its own economic impact on land—property taxes, income taxes, and estate taxes.

Property taxes are the major source of revenue for most local governments. Since land is a more critical resource to agriculture than it is in most other industries, agriculture is particularly vulnerable to the burden of property taxes. In many situations these taxes may be transferring most of the land rents from the landowners to local government. This is not the place to debate the efficiency or equity of such policies. It is apparent, however, that if property taxes take a higher proportion of land rents in agricultural uses than in alternative uses, the competitive position of agriculture for land acreage will be weakened.

Because of perceived inequities in applying the property tax to farmland and a desire to influence land use, by May 1974, 34 states had adopted some form of differential tax assessment. There are three general types of differential taxation for agriculture—straight preferential assessment, deferred taxes, and restrictive agreements.

In 1974, nine states had straight preferential assessment. (The land is valued at its current use rather than at its market value with no penalty for later conversion to another use.)

A second and the most common of the differential assessments is the deferred tax. The property is taxed at its current use; but when that use changes, there is a "rollback" tax on the difference in assessed value between its agricultural use and its higher valued use for a specified period. The New Jersey Farmland Assessment Act of 1964 is an example. Once the use changes, the tax rate is applied on the difference between the special assessment and the full market value for that year and for the previous two years.

The restrictive agreement is an arrangement in which the landowner and the local government contract to restrict land use in exchange for a preferential assessment. In 1974, ten states utilized this arrangement (six of them used the contract system for some of their land and either the preferential or deferred system for other lands). The California Land Conservation Act (CLCA) of 1965 offers cities and counties the option of making legally binding 10-year contracts with landowners whose land meets certain open-space or agricultural criteria. Unless notification of nonrenewal is given (by either party), the contract is automatically extended each year. Cancellation due to the opportunity for a higher return, or due to an unprofitable farming operation, is not allowed. Upon notice of nonrenewal, the contract extends another nine years. The assessed value increases to 60 percent of its market value immediately and continues to full value at cancellation. There is also a penalty, unless waived, of 12.5 percent of the full market value of the land. Indeed, the cancellation procedure is so complex and the penalty so onerous that landowners who want to keep their options open are discouraged from enrolling in the program.

In 1971, 9.5 million acres (3.8 million hectares) of California land were in the differential assessment program—about one-fifth of all the private land in the state. But, as Carman and Polson (1971) noted, early experience showed that "land placed in the program was below average value and probably in little immediate danger of being converted to nonagricultural use." In a detailed survey study of three sections of Sacramento County, it was found that the highest enrollment in CLCA was in the foothill section and thus the least exposed to urban development (a good description of the act and an analysis of its effectiveness are found in Hansen and Schwartz, 1975). The evidence seems to suggest that the CLCA—although one of the strongest land-use acts—is basically a subsidy to agriculture and is ineffective in curbing urban sprawl.

A second tax having some effect on farm real estate values is the income tax. Investment in agricultural land has long been an effective tax shelter to the high-income investor seeking to convert ordinary income to capital gains. The fact that earnings from the realized appreciation in real estate are taxed as capital gains rather than ordinary income has increased the demand for land held for speculative purposes. Landowners on the urban fringe are in a position to capture these large capital gains, and the typical "rollback" penalty of the differential assessment schemes is insufficient to counteract the advantages of conversion and keep the land in agriculture. And the blow of a possible penalty is also lessened because state and local taxes are deductible from federal taxes.

A third major category of taxation that affects real estate and farmlands is the estate tax. Many farmers have an income problem when they retire. Since they are not eligible for many public retirement programs, they may be dependent on their wealth in the form of appreciated land values. Also, at the time of transfer to the next generation, there is an urgent need for liquid assets to pay the estate taxes. Especially if the property taxes are based on the full market value of the land, there may be pressure for land conversion and sale. The Tax Reform Act of 1976, allowing a much larger exemption before estates are taxed and other advantages, may alleviate this situation somewhat.

Environmental Energy and Labor Policies

Two federal environmental statutes with great potential impact on agriculture are the Federal Environmental Pesticide Control Act of 1972 (Public Law 92-516) and the Federal Water Pollution Control Act Amendments of 1972 (Public Law 92-500). The first required all pesticides to be registered with the Environmental Protection Agency (EPA), which would regulate their use. The EPA may inspect pesticide manufacturers and can issue stop-sale and seizure orders. It may also order manufacturers to make indemnity payments for pesticides that cause environmental and human health damage.

An illustration of the potential economic impact on farmers of Public Law 92-516 is in the restriction or prohibition of important herbicides. Nalewaja (1975) shows that the elimination of herbicides on corn alone would require 17.7 million people to weed by hand to obtain the same level of weed control achieved by herbicides.

DDT and other insecticides have been banned, but it is difficult to determine the effect on agriculture. Perhaps even more important than past bans is the threat of future ones and the increasing regimentation of the entire industry by the government.

Public Law 92-500 has some ominous features if rigidly enforced on agriculture. The law represents a major change in the approach to water pollution control. The potentially stringent regulations can be understood when viewed in the light of the overall goal—to eliminate all pollutant discharges into the water in the United States by 1985.

The impact of Public Law 92-500 on irrigated agriculture can be seen in the dilemma of the Colorado River basin, where salinity is the major water-quality problem. The salt flow must be controlled in the upper basin if high-quality water is to be delivered to the thirsty lower basin, but the two major methods of control—improved irrigation efficiency and return flow treatment—are costly. In addition, many upper basin areas are characterized by small farms, low capitalization, and many residents with incomes at or below the poverty level. Thus the burden would be great indeed if these farmers were required to bear the cost of water-quality improvement.

Modern farming in the United States has become highly dependent on energy from fossil fuels. An energy crisis resulting from embargoes by the Organization of Petroleum Exporting Countries (OPEC) or resource exhaustion would bring about major changes in agricultural production and in world food prices.

Dvoskin and Heady (1976), using a large linear programming model of the agricultural sector, analyzed potential changes in agriculture resulting from changes in the amount of energy available, in the prices paid for energy, and in increased export demand for food along with higher energy prices. Constraining the energy consumption to 90 percent of its 1972 level caused a 55 percent increase in farm output prices, whereas doubling the energy prices from their 1974 level resulted in only a 13 percent increase in farm prices. The analysis showed that a proportional decrease of energy use in all input categories is not the least-cost alternative. The most energy-saving strategy was a 41 percent reduction in the acres of irrigated land. The far-reaching implications of this sort of change would lead to major shifts among regions of production, radical modifications of crop mixes, and redistribution of farm income. The second largest change was in the reduction in the purchase of commercial nitrogen and its replacement by nitrogen from organic sources.

The relation between labor policies and land use is quite indirect. Still, labor policies may have a differential impact on the profitability of agricultural production in certain regions and thus affect the land area devoted to agriculture.

The Occupational Safety and Health Act of 1970 (Public Law 91-596) established a comprehensive on-the-job safety program for about 55 million industrial, farm, and construction workers. Employers are required to furnish a workplace free from "recognized hazards." Penalties for noncompliance may run up to $10,000 or six months in prison or both. Since agriculture has a poor job-safety record, the potential costs for full compliance are substantial.

Agricultural workers have historically been excluded from most federal and state labor laws. But there has been a recent trend toward extending the benefits enjoyed by workers in other industries to those in agriculture. These benefits may come to farm workers from several sources—federal and state government, workers' unions, and the employer. The economic impact on growers of this increasingly liberal treatment of farm workers, however, cannot be immediately foreseen. Higher costs may be more than offset by greater worker productivity.

At the federal level, the Fair Labor Standards Act of 1938 was amended in 1966 to give farm workers the benefit of minimum-wage and overtime-pay legislation. In a 1974 amendment, a few of the many exemptions to the coverage were eliminated. Local commuting pieceworkers are now counted in the computation of the 500 man-days required for employer participation, and the partial overtime pay exemptions are eliminated. But there are still many exclusions to the coverage; even where a minimum wage is established, it is often lower than in other industries.

A second major change at the federal level occurred January 1, 1977, with the extension of unemployment compensation to farm workers whose employers hired four or more workers in each of 20 weeks in a year or paid at least $5,000 in wages in a calendar year. Excluded from coverage until January 1, 1979, are nonresident aliens admitted to the United States to do farm work.

Individual states may go beyond the federal government in extending benefits to the agricultural workers. They may, for example, have a higher minimum wage. Workmen's Compensation is basically a state system. But farm workers constitute the largest group excluded by many of the states from coverage. Only 16 states have some compulsory coverage for farm workers. In Alabama and in Texas, farm worker coverage is specifically prohibited.

California has been one of the most progressive states in extending benefits to agricultural workers. The average wage rate for field and livestock workers in California is nearly 15 percent above the national average. As a result of the Agricultural Labor Relations Act of 1975, unionization of farm workers will gradually encompass more California farms, large and small. A typical United Farm Workers or Teamsters contract includes benefits that compare favorably with those enjoyed by workers in other industries. Such amenities as pay for overtime or standby, rest periods, paid vacations, bereavement leave, holidays, time off for jury or witness duty, travel, health insurance, health and safety provisions, and pensions may be and now often are included in the contract.

FOREST LAND UTILIZATION

There are a number of parallels and interrelationships between agricultural lands and forest lands. In fact, slightly more than one-fourth of the 500 million acres (about 202 million hectares) of commercial forest land in the nation is included in farm holdings.[2] The public policy toward forest land has developed in a significantly different way from the policy for agricultural land. Nonetheless, in recent years the general effects of policy developments on the use of land for food and fiber production have been much the same for both agricultural and forest land.

The importance of using land for the production of food has been recognized since colonial times. In contrast, the forest, although clearly a useful source of wood for fuel and forest products, was seen largely as a barrier to agricultural development. Indeed, it has been estimated that on about three-fourths of the forest area laboriously cleared for agriculture the wood was simply burned in place without any use being made of it. The use of forest land as the site for growing harvestable crops of timber, rather than simply as a storehouse of timber for liquidation, is largely a development of the 20th century.

Public Forest Land

The turning point in the long historic process of transferring public domain lands to private and state ownership is often said to have been in 1891, with the enactment of a law (16 U.S.C. 471) which authorized the president to set aside forest reserves and which marked the first systematic provision for withholding a part of the public domain from disposal. This act of 1891 was followed by a long series of reservations and purchases of national forests, national parks and related areas, wildlife refuges, and military reservations, with both the disposal and the initial reservation phases being finally brought to a close in 1976 by passage of the Federal Land Policy and Management Act (Public Law 94-579), which blanketed in the remaining unreserved public domain as National Resource Lands. Through this process about one-third of the nation's land has been reserved in federal ownership. This federal domain includes 187 million acres (75.7 million hectares) in the National Forest System, of which 92 million acres (37.2 million hectares) are commercial forest land. Other federal holdings include another 15 million acres (6.1 million hectares) of commercial forest land, giving a total of 107 million acres (43.3 million hectares) or about 21 percent of the nation's total commercial forest area.

The purpose of the early forest reserves (which, in time, became the National Forest System) was not established until the Organic Act of 1897 (16 U.S.C. 473), which specified that these reserves were established "for the purpose of securing favorable conditions of waterflow, and to furnish a continuous supply of timber for the use and necessities of citizens of the United States." Although timber production was a primary reason for setting aside the National Forests, these areas were managed largely on a custodial basis with relatively little development until the end of World War II. At that time, the combination of high consumer demand and the depletion of private timber holdings resulted in a rapid increase in the cut from the National Forests, with the volume rising from the 1940 level of 2 billion board feet (4.7 million cubic meters) to nearly 14 billion board feet (33 million cubic meters) in 1970 and 1972. Subsequently, the cut has dropped to 11 billion board feet (26 million cubic meters) in 1976.

The demand for other uses of forest land has also increased rapidly. In 1960 the already well-established, multiple-use management policy for the National Forests was given full legal expression by adoption of the Multiple Use Sustained Yield Act (16 U.S.C. 529), which specified that National Forests "are established and shall be administered for outdoor recreation, range, timber, watershed, and wildlife and fish purposes." No priorities among these purposes were indicated.

At the same time, a number of actions were taken which amounted to a second round of reservations on the already reserved federal lands. Through measures such as the establishment of National Recreation Areas and the adoption of laws including the Wilderness Act of 1964 (16 U.S.C. 1131), the National Trails System Act of 1968 (16 U.S.C. 1241), the Wild and Scenic Rivers Act of 1968 (16 U.S.C. 1271), and the Endangered Species Act of 1973 (Public Law 93-205), Congress has established dominant uses and special constraints on the development for substantial areas of National Forest and other federal forest land. There has been no similar designation of areas to be used primarily for timber production. Instead, when the Public Land Law Review Commission in 1970 recommended the establishment of dominant timber production units, there was a highly vocal protest against the concept of dominant use (except for recreation, wilderness, or wildlife) and the commission's report generally.

In addition, the Forest Service itself, in its land-use planning, has established various special-use zones, including roadside and streamside areas and other special component lands where timber management must be sharply constrained on the basis of environmental, aesthetic, or other considerations. Thus, it now appears that something less than 60 percent of the presumed 92 million acres (37.2 million hectares) of commercial forest land in the National Forest System can be considered as standard component land, which is the basis for determining the allowable timber cut.

To this can be added the effects of the wave of environmental legislation of the last decade, notably the National Environmental Policy Act of 1969 (Public Law 91-190), the Clean Air Act Amendments of 1970 (Public Law 91-604), and the Federal Water Pollution Control Act Amendments of 1972 (Public Law 92-500). The fact that the section of the Code of Federal Regulations entitled "Protection of the Environment" increased from 460 pages in the 1972 edition to almost 2,000 pages in the 1974 edition is indicative of the rapidly growing complexity and pervasiveness of the system of public controls in this area.

Finally, there are the still uncertain effects of the National Forest Management Act of 1976 (Public Law 94-588), enacted in the final days of the congressional session after a long and at times bitter debate. The act is lengthy, complex, and somewhat ambiguous. Although it is less restrictive than the Randolph bill, which was favored by preservationist groups among others, on balance it appears substantially more constraining of timber management than the policies in effect before a court interpretation of a portion of the act of 1897 precipitated a timber crisis which led in time to this new legislation. There are provisions in the act which could enable substantial flexibility in timber management and harvesting, but these are largely in the form of permitted exceptions to general standards established in the act. A full evaluation of the impact of the act will not be possible until the implementing regulations have been developed and some operating experience under these new provisions has been gained.

No studies have been conducted which show all the effects of these policies on the use of federal forests for the production of wood. In the nature of the situation, it seems unlikely that such quantitative estimates could be made. Certainly, the net effect of these policies is an appreciable reduction of the role of these forests in growing wood and contributing to the nation's timber supply. Under the impact of these policies and continuing political pressures, the cut from the National Forests seems far more likely to be held to the level of 12 billion board feet (28.3 million cubic meters) or less rather than being increased to the more than 18 billion board feet (42.5 million cubic meters) which, until recently, the U.S. Forest Service has indicated would be justified under intensive timber management.

The tendency of public policy has been to weight environmental effects and other forest uses far more heavily relative to timber output than in the past. The result, whether intended or not, has been to constrain and reduce the possibilities for wood production. There has been much talk of the evaluation of trade-offs and similar matters. In the absence of any clear policy on the role of the federal forests in wood production, or of any timber output targets for these forests, there can be little basis for evaluating what is being traded.

It should be mentioned that the Forest and Rangeland Renewable Resources Planning Act of 1974 (Public Law 93-378) did establish a systematic process for developing quantitative output goals for the National Forests and developing programs for their achievement. Amendments to this act, included in the National Forest Management Act of 1976, were three times as long as the 1974 act itself. Perhaps all that should be said at this point—and certainly the most optimistic statement which can be justified—is that it is still too early to judge the possible effectiveness of this act.

Private Forest Land

The 79 percent of the commercial forest land which is not in federal ownership is largely privately held. This nonfederal forest land includes 29 million acres (11.7 million hectares) held by state and local governments, 67 million acres (27.1 million hectares) owned by firms engaged in the production of forest products, 131 million acres (53 million hectares) held on farms, and 165 million acres (66.8 million hectares) held in miscellaneous private ownerships (typically small in size). Public policy has a different effect on the role of these lands in wood production than it does in relation to federal forests. There is, of course, wide variation among the 50 states in this regard. However, at least in the states that are the leaders in timber cut and timber growth, there are a number of policies and institutions clearly designed to encourage the allocation of land to wood production.

At least 16 states have laws regulating forest practices on private forest lands. The more recent of these—the 1971 Forest Practice Act of Oregon and the 1973 act of California—have strong requirements to ensure adequate restocking after logging and to protect the productivity of the land for forest growth. In California there are also definite legal constraints on the conversion of forest land to other uses.

Another policy area in which the states have moved widely to encourage the use of land for wood production is in property taxation. It has long been argued that since the production cycle for forest stands is a multiple of 20 to 100 or more of the annual tax cycle, the imposition of the general property tax on forest stands involves a time bias which discourages the use of land for wood production. In response to this argument, at least 34 states have adopted a variety of special forest tax laws as alternatives to or replacements for the general property tax. All of these laws are intended to facilitate and encourage the holding of land in timber growing, although their effectiveness appears highly variable.

The 1976 California timber yield tax law, which became effective in 1977, is undoubtedly the most far-reaching of these laws. It imposes a mandatory yield tax on all trees cut for forest products in the state and provides for the establishment of timberland preserve zones through publicly initiated action; the timber will be exempt from the general property tax and the land will be assessed for general property tax purposes on the basis of its use in growing timber and other prescribed compatible purposes. Land so classified is subject to enforceable restrictions on use for a period of 10 years. This period is automatically extended each year unless notice of rezoning is given. While this law, in intent and effect, is favorable to the use of the classified land for timber production, such classification will largely remove the land from the influence of market forces in the determination of land use.

In terms of federal policy, special provisions in the Internal Revenue Code, which enable revenues resulting from increases in timber values to be classified as capital gains under specified conditions, greatly encourage investment in timber growing, although it is often argued that these benefits accrue largely to the larger corporate owners of forest land. Similarly, a number of federal subsidies for timber growing have been developed over the years: the Forestry Incentives Act of 1973 is the most recent. These subsidy programs are generally administered in cooperation with the states and are usually oriented toward the small forest owners.

Although such federal programs work to favor the use of forest land for timber growing, indirectly public policy involving the federal forests has the opposite effect. Here it should be noted that although the federal forests contain only 21 percent of the commercial forest land, the National Forest System has 51 percent of the national softwood sawtimber volume; other public forests, both federal and state, have an additional 12 percent. Thus, policies which sharply constrain the level of cut on these federal forests and emphasize stability in physical terms result in heavy pressures to liquidate timber on private lands to meet current demands and contribute significantly to a price instability which is discouraging to private investment in timber growing.

The wave of environmental legislation mentioned earlier seems certain to have an important effect on the use of land for timber production, although many of the key provisions are only now being brought to bear on private forest lands. Although all of the major environmental acts are involved, Section 208 of the Federal Water Pollution Control Act Amendments of 1972 may be the most significant single provision; it requires each state to develop areawide plans for control of water pollution from both point and nonpoint sources. The regulations developed under Section 208 encourage the imposition of land-use zoning controls and, in the case of nonpoint sources, require an evaluation of measures necessary to produce the desired level of control by applying best management practices. These requirements could bring about a rapid extension of state regulation of forest practices—possibly under federal guidelines—and are resulting in confusion and conflict between state water-quality agencies and forestry agencies over operational responsibility for the control of private forestry practices. Other potential impacts include the possibility that the use of prescribed fire as a forest management tool may be greatly restricted by regulations designed to protect air quality, and the possibility that the use of herbicides and other chemicals in forest management may be restricted under the requirements for registration of chemicals by the EPA under Public Law 92-516 and other legislation controlling the use of toxic substances.

State environmental laws—usually modeled after federal laws—are also important, and direct action at the county level is increasingly common.

All of this is placing constraints on land use which increase the cost of management. The landowner must obtain clearances from numerous agencies—with uncoordinated and sometimes conflicting roles—in order to carry out any form of management. In addition, there is an unsettling sense of uncertainty about what uses are permitted and what the actual rights are in landownership. Such an atmosphere of uncertainty inhibits efficient land use, particularly in regard to the long-term investments required for timber growing and wood production.

CONCLUSION: THE DECLINING ROLE OF MARKET FORCES

As the preceding sections amply demonstrate, there is increasing public intervention, through statutes and regulatory policies, which affects production decisions and income distribution in agriculture and forestry.

The public recognition that many valuable nonfood amenities also flow from agricultural and forest land has brought a spate of policies that seriously impinge on private property rights in land and the user surplus derivable from employing land in agricultural and forest production. The result is that although in a physical sense there is plenty of acreage to produce our food and fiber for many decades, in an economic sense it is by no means clear that it will be profitable to do so. If not, our society has at least three options: (1) Land will be removed from agricultural and forest use as resource owners search elsewhere for larger user surpluses. This will continue until food and fiber supplies are reduced, prices rise, user surpluses rise again to competitive levels, and an equilibrium is established. (2) The more onerous policies that reduce user surpluses in agriculture and forestry will be repealed or softened to lessen their negative effects. (3) Coercive measures will be used by the regulatory agencies to keep resources in agriculture at below competitive user surpluses. The choice will not be easy, since many considerations must be brought to bear on the political decision. The authors of this chapter, however, believe that the third option is likely to be most inefficient as well as generally most inequitable; yet at present it seems to be the most popular.

In the case of forests and public domain lands, by the beginning of the 20th century public policy was becoming oriented toward holding portions of the public domain in federal ownership for timber production and livestock grazing rather than allowing the transfer of title to private ownership and the full play of market forces, with their perceived imperfections. But the products taken from the public lands (timber and livestock) were recognized as market goods to be allocated to the highest bidder at the appropriate time. As this early public purpose of providing timber supplies and livestock grazing was realized through a rapid increase in the federal timber cut following World War II (but not in permitted livestock numbers), public policy increasingly responded to intensified demand for other values of the public lands which were thought to conflict with timber and livestock yields at the demand levels involved. This was followed by a policy response to external costs of forest and grazing operations in the form of environmental effects. Thus, public policies responsive to extramarket uses of the public lands—to public goods and to externalities—in general have served to diminish rapidly the already restricted role of market forces in determining the use of the public lands and to constrain sharply their role in timber and livestock production.

For the 73 percent of the commercial forest area in private ownership, a reliance chiefly on market forces for allocation has continued until recent years. Public policy has reinforced and intensified the market forces favoring an allocation of resources to timber production through special tax provisions and through subsidies for timber-growing investments. In addition, in a minority of the states, the police power has been invoked to regulate forest practices and require the maintenance of timber productivity in areas subject to logging. In California, such policies have been developed to the extent that they will largely insulate the main nonfederal forest areas from the effects of future market changes and maintain them as open space for timber growing.

Implementation of the major environmental and safety legislation enacted since 1969 is now beginning to have a direct effect on these private forests as well as private agricultural lands generally. This greatly increased policy recognition of externalities and social interests in the use of private land seems certain to constrain decision making by the owners to a major degree, as well as impose additional costs and complexities in the management of this land for timber and agriculture and reduce substantially the role of market forces in determining the use of the land. The result is likely to be a reduction of the role of this land in food and fiber production and quite possibly an increase in the tendency to hold such land (especially in small holdings) in an inactive and unmanaged condition. Reserving so-called prime land for agricultural use is perhaps the most blatant societal effort to remove the market from any allocative function in land use.

The forces leading to this rapidly diminishing role of the market in determining the allocation of agricultural and forest land resources reflect changing social values. No judgment can be made on the basis of this analysis about the effectiveness or desirability of these changes. However, in shifting from reliance on the market to an increasing reliance on political processes, society has failed to establish comprehensive allocation criteria to replace those that are being abandoned.

Acknowledgment

The authors acknowledge with gratitude the assistance of Carole Nuckton in reviewing the literature, making computations, and editing.

Notes

1. Perhaps the best study of the phenomenon of the postwar rise in the real value of agricultural land is by J. E. Reynolds and J. F. Timmons (1969). Independent variables regressed on a time series of agricultural land values, and the signs of the regression coefficients were as follows: (1) expected net farm income (positive), (2) government farm program payments (positive), (3) expected capital gains (positive), (4) technological advance (positive), (5) farm enlargement (positive), (6) the number of voluntary transfers of farmland (negative), and (7) an increasing demand for land from a growing population (positive).

2. Under the National Forest Survey, forest land capable of growing in excess of 20 cubic feet per acre per year of industrial wood in natural stands and not legally withdrawn from timber utilization is classified as commercial forest land. No consideration is given to the actual growth in the area, the intent of the owner, the accessibility of the land, or the financial feasibility of growing timber on the land on a commercial basis. Thus, the reported 500 million acres of commercial forest land greatly overstates the area now being used, or likely to be used, for the growing of timber on a commercial basis.

References

Carman, Hoy F., and James G. Polson. 1971. Tax shifts occurring as a result of differential assessment of farmland: California, 1968–69. *National Tax Journal* 24(4):449–457.

Cotner, Melvin L. 1976. Land use policy and agriculture: A national perspective. U.S. Dep. Agr. Econ. Res. Serv. No. 630.

Dvoskin, Dan, and Earl O. Heady. 1976. U.S. agricultural production under limited energy supplies, high energy prices, and expanding agricultural exports. Center for Agricultural and Rural Development, Iowa State University, Report No. 69.

Hansen, David E., and Seymour I. Schwartz. 1975. Landowner behavior at the rural-urban fringe in response to preferential property taxation. *Land Economics* 51(4):341–354.

Krause, Orville, and Dwight Hair. 1975. Perspectives on prime lands. Background papers for Seminar on Retention of Prime Lands. U.S. Department of Agriculture, Washington, D.C.

Nalewaja, J. D. 1975. Energy requirement of various weed control practices, North Dakota State University. Quoted in Dan Dvoskin and Earl O. Heady. 1976.

Reynolds, John E., and John F. Timmons. 1969. Factors affecting farmland values in the United States. Iowa State Agr. and Home Econ. Exp. Sta. Res. Bull. 566.

Tatom, John A. 1976. The welfare cost of inflation. Federal Reserve Bank of St. Louis, *Review* 58:11.

Contributors

JAMES S. BETHEL is dean of the College of Forest Resources at the University of Washington.

EMERY N. CASTLE is senior fellow and vice president, Resources for the Future, Inc., Washington, D.C.

ERIC L. ELLWOOD is dean of the School of Forest Resources, assistant director of the North Carolina Agricultural Experiment Station in the School of Agriculture and Life Sciences, and professor of wood and paper science, North Carolina State University, Raleigh.

B. DELWORTH GARDNER is director of the Giannini Foundation, University of California, Berkeley, and professor of agricultural economics, University of California, Davis.

MARTIN A. MASSENGALE is vice chancellor at the University of Nebraska, Lincoln, and member of the Institute of Agriculture and Natural Resources.

WILLIAM E. SPLINTER is head and professor of agricultural engineering, Department of Agricultural Engineering, Institute of Agriculture and Natural Resources, University of Nebraska, Lincoln.

GEORGE R. STAEBLER is director of forestry research for the Weyerhaeuser Company, Centralia, Washington.

ALISON C. THORNE is lecturer, Departments of Sociology and Home Economics, Utah State University, Logan.

WYNNE THORNE is professor emeritus, Department of Soils and Biometeorology, Utah State University, Logan.

SYLVAN H. WITTWER is director of the Agricultural Experiment Station, assistant dean of the College of Agriculture and Natural Resources, and professor of horticulture, Michigan State University, East Lansing.

JOHN A. ZIVNUSKA is professor of forestry, University of California, Berkeley.

Index